Applied Atmospheric Dynamics

Applied Atmospheric Dynamics

Amanda H. Lynch
Monash University

John J. Cassano
The University of Colorado

John Wiley & Sons, Ltd

Copyright © 2006   John Wiley & Sons Ltd, The Atrium, Southern Gate, Chichester,
West Sussex PO19 8SQ, England

Telephone (+44) 1243 779777

Email (for orders and customer service enquiries): cs-books@wiley.co.uk
Visit our Home Page on www.wiley.com

All Rights Reserved. No part of this publication may be reproduced, stored in a retrieval system or transmitted in any form or by any means, electronic, mechanical, photocopying, recording, scanning or otherwise, except under the terms of the Copyright, Designs and Patents Act 1988 or under the terms of a licence issued by the Copyright Licensing Agency Ltd, 90 Tottenham Court Road, London W1T 4LP, UK, without the permission in writing of the Publisher. Requests to the Publisher should be addressed to the Permissions Department, John Wiley & Sons Ltd, The Atrium, Southern Gate, Chichester, West Sussex PO19 8SQ, England, or emailed to permreq@wiley.co.uk, or faxed to (+44) 1243 770620.

Designations used by companies to distinguish their products are often claimed as trademarks. All brand names and product names used in this book are trade names, service marks, trademarks or registered trademarks of their respective owners. The Publisher is not associated with any product or vendor mentioned in this book.

This publication is designed to provide accurate and authoritative information in regard to the subject matter covered. It is sold on the understanding that the Publisher is not engaged in rendering professional services. If professional advice or other expert assistance is required, the services of a competent professional should be sought.

Other Wiley Editorial Offices

John Wiley & Sons Inc., 111 River Street, Hoboken, NJ 07030, USA

Jossey-Bass, 989 Market Street, San Francisco, CA 94103-1741, USA

Wiley-VCH Verlag GmbH, Boschstr. 12, D-69469 Weinheim, Germany

John Wiley & Sons Australia Ltd, 42 McDougall Street, Milton, Queensland 4064, Australia

John Wiley & Sons (Asia) Pte Ltd, 2 Clementi Loop #02-01, Jin Xing Distripark, Singapore 129809

John Wiley & Sons Canada Ltd, 6045 Freemont Blvd, Mississauga, Ontario, Canada L5R 4J3

Wiley also publishes its books in a variety of electronic formats. Some content that appears in print may not be available in electronic books.

Library of Congress Cataloging in Publication Data

Lynch, Amanda H.
  Applied atmospheric dynamics / Amanda H. Lynch and John J. Cassano.
    p. cm.
  Includes bibliographical references and index.
  ISBN-13: 978-0-470-86172-1 (acid-free paper)
  ISBN-10: 0-470-86172-X (acid-free paper)
  ISBN-13: 978-0-470-86173-8 (pbk.: acid-free paper)
  ISBN-10: 0-470-86173-8 (pbk.: acid-free paper)
  1. Atmospheric physics. 2. Dynamics—Mathematics. I. Cassano, John J. II. Title.

QC861.3.L957 2006
551.5—dc22
                                                                        2006016133

British Library Cataloguing in Publication Data

A catalogue record for this book is available from the British Library

ISBN-13: 978-0-470-86172-1 (HB)   978-0-470-86173-8 (PB)
ISBN-10: 0-470-86172-X (HB)   0-470-86173-8 (PB)

Typeset in 10.5/12.5pt Times by Integra Software Services Pvt. Ltd, Pondicherry, India
Printed and bound in Great Britain by Antony Rowe Ltd, Chippenham, Wiltshire
This book is printed on acid-free paper responsibly manufactured from sustainable forestry in which at least two trees are planted for each one used for paper production.

Contents

Preface	ix

Part I Anatomy of a cyclone — 1

1 Anatomy of a cyclone — 3
- 1.1 A 'typical' extra-tropical cyclone — 3
- 1.2 Describing the atmosphere — 4
- 1.3 Air masses and fronts — 9
- 1.4 The structure of a typical extra-tropical cyclone — 14
- Review questions — 20

2 Mathematical methods in fluid dynamics — 23
- 2.1 Scalars and vectors — 23
- 2.2 The algebra of vectors — 23
- 2.3 Scalar and vector fields — 27
- 2.4 Coordinate systems on the Earth — 27
- 2.5 Gradients of vectors — 28
- 2.6 Line and surface integrals — 31
- 2.7 Eulerian and Lagrangian frames of reference — 34
- 2.8 Advection — 35
- Review questions — 38

3 Properties of fluids — 41
- 3.1 Solids, liquids, and gases — 41
- 3.2 Thermodynamic properties of air — 42
- 3.3 Composition of the atmosphere — 43
- 3.4 Static stability — 46
- 3.5 The continuum hypothesis — 50
- 3.6 Practical assumptions — 50
- 3.7 Continuity equation — 51
- Review questions — 53

4 Fundamental forces — 57
- 4.1 Newton's second law: $F = ma$ — 57
- 4.2 Body, surface, and line forces — 57

	4.3	Forces in an inertial reference frame	58
	4.4	Forces in a rotating reference frame	66
	4.5	The Navier–Stokes equations	72
		Review questions	74

5 Scale analysis 79

5.1 Dimensional homogeneity	79
5.2 Scales	80
5.3 Non-dimensional parameters	80
5.4 Scale analysis	84
5.5 The geostrophic approximation	87
Review questions	91

6 Simple steady motion 93

6.1 Natural coordinate system	93
6.2 Balanced flow	95
6.3 The Boussinesq approximation	104
6.4 The thermal wind	105
6.5 Departures from balance	108
Review questions	114

7 Circulation and vorticity 119

7.1 Circulation	119
7.2 Vorticity	124
7.3 Conservation of potential vorticity	126
7.4 An introduction to the vorticity equation	130
Review questions	132

8 Simple wave motions 135

8.1 Properties of waves	135
8.2 Perturbation analysis	138
8.3 Planetary waves	140
Review questions	147

9 Extra-tropical weather systems 149

9.1 Fronts	149
9.2 Frontal cyclones	152
9.3 Baroclinic instability	161
Review questions	162

Part II Atmospheric phenomena 165

10 Boundary layers 167

10.1 Turbulence	168
10.2 Reynolds decomposition	169
10.3 Generation of turbulence	172
10.4 Closure assumptions	173
Review questions	181

11 Clouds and severe weather — **183**
- 11.1 Moist processes in the atmosphere — 183
- 11.2 Air mass thunderstorms — 191
- 11.3 Multi-cell thunderstorms — 193
- 11.4 Supercell thunderstorms and tornadoes — 194
- 11.5 Mesoscale convective systems — 196
- Review questions — 197

12 Tropical weather — **199**
- 12.1 Scales of motion — 199
- 12.2 Atmospheric oscillations — 203
- 12.3 Tropical cyclones — 205
- Review questions — 208

13 Mountain weather — **209**
- 13.1 Internal gravity waves — 209
- 13.2 Flow over mountains — 217
- 13.3 Downslope windstorms — 226
- Review questions — 230

14 Polar weather — **233**
- 14.1 Katabatic winds — 233
- 14.2 Barrier winds — 242
- 14.3 Polar lows — 247
- Review questions — 249

15 Epilogue: the general circulation — **251**
- 15.1 Fueled by the Sun — 251
- 15.2 Radiative–convective equilibrium — 253
- 15.3 The zonal mean circulation — 254
- 15.4 The angular momentum budget — 259
- 15.5 The energy cycle — 261

Appendix A – symbols — **265**

Appendix B – constants and units — **271**

Bibliography — **273**

Index — **277**

Preface

Many can brook the weather that love not the wind.
W. Shakespeare, *"Love's Labour's Lost"*, Act IV, Scene II

During mid-February 2003, one of the biggest winter storms on record cut a swathe through the mid-western and eastern United States. Low temperature and snowfall records were set all along the eastern seaboard. Tornadoes, extreme hail, flooding, and mudslides were all generated by the 'beast in the east'. When all was said and done, 45 people lost their lives, and 122 people were injured, as a direct result of the weather.

The weather can be a cause of disruption, despair, and even danger everywhere around the world at one time or another. Even when benign, it is a source of constant fascination for many people. Yet connecting this interest with the underpinnings of fluid mechanics has remained beyond reach for many. It is our hope with this book to make the intriguing ways in which the atmosphere moves accessible to a broader range of students and general readers. We have done this by linking real physical events with theoretical models at every possible juncture. The storm of February 2003 provides a valuable illustration of many of the important concepts in atmospheric dynamics, and we have used many other dramatic weather events as well, from the devastating Hurricane Katrina to the strong katabatic flows of Antarctica. The level of mathematics required, though not rudimentary, is pre-vector calculus, and the emphasis is always on the phenomenology.

Part I takes the reader through all of the basic concepts required to understand the development and decay of mid-latitude low-pressure systems. These concepts include balanced and unbalanced flows, vorticity, and waves. In Part II, a broader range of phenomena are considered, ranging from the tropics to the poles. These later chapters can be considered in any order. For each of the 14 chapters, review questions to test understanding and to provide practice are posed, the worked solutions of which are available. The book ends with a discussion of the role of weather systems in maintaining the global circulation.

The accompanying CD-ROM includes all of the illustrations in the book in JPEG format, and many more besides. Animations and videos of important processes, satellite pictures of interesting events, and weather maps of all varieties are collected on the CD-ROM. All of the available data for the storm of February 2003 is in a searchable database, suitable for a range of investigations. In addition, four possible research projects are included, on Atlantic hurricanes, southern hemisphere cyclones, polar lows, and tornadoes.

While only two authors are listed on the front of this book, many others contributed. Elizabeth Cassano stepped up whenever asked, to prepare content for the CD-ROM, to frame project material, and to find that last data point. Christopher Takeuchi, Mark Seefeldt, and Petteri Uotila provided a number of figures for the book and CD-ROM, which Henry Johnson and Casey Tonkin tested to breaking point. Michael Shaw provided invaluable assistance in proofreading the book from a student's perspective and solving all of the review questions. David Underwood created a number of figures and animations for the book and CD-ROM. Barbara Lynch's advice, scientific and literary, was priceless.

AHL and JJC

I want to thank the many students of atmospheric dynamics that I have taught over the years – without their enthusiasm, excellence, and well-targeted criticism, this book would not have been written. They also served as guinea pigs for many of the approaches we have used here. I also wish to thank my parents, for their boundless faith in me; my husband, for his most practical support; and my daughters, for bringing me down from the clouds.

AHL, Melbourne, Australia

I must also thank my family for their unending support in all of my life's endeavors, and in particular my parents Emilia and Vito for giving me the opportunity to pursue an educational and professional path that perhaps was not the one they had hoped I would follow. Finally, I must thank my wife Elizabeth for her love, support, and patience. Not only did she help with numerous tasks related to the writing of this book, she also provided me with the perspective to realize that there is more to life than work and weather.

JJC, Boulder, Colorado
November 2005

Part I Anatomy of a cyclone

1 Anatomy of a cyclone

1.1 A 'typical' extra-tropical cyclone

A snow emergency was declared in Boston, Massachusetts as a record snowfall paralyzed the city. Airports in Washington, DC and New York City were closed, and trains and buses were cancelled. The blizzard, dubbed 'the beast in the east' by the media, dumped heavy snow in a broad swathe from Iowa to New England on the President's Day holiday weekend of 2003 (15–17 February).

This storm had been traversing the United States for a week – leaving disaster in its wake in some places, and hardly being noticed in others. Flooding and mudslides occurred in southern California. Ten tornadoes were reported in the south-eastern United States. In Delaware, over 500 people were evacuated from a townhouse complex after water from melting snow began leaking into the electrical system. Meanwhile, hail and strong winds were reported from the southern Plains to the south-eastern United States, downing trees and power lines and removing roofs. During the period that this storm crossed the country, 45 people were killed and 122 people injured as a direct result of the weather (Angel 2003). Over US$144 million in damage was reported.

These diverse weather events were the result of a low-pressure system outside the tropics, an *extra-tropical cyclone*, that crossed the United States from 11 to 19 February 2003.

The system first approached the southern California coast on 11 February. Over the next two days it crossed the mountainous western portion of the United States, looking relatively weak and sometimes disappearing from surface maps. Then, on 13 February, the surface low-pressure center redeveloped over south-eastern Colorado, as such systems often do when crossing the Rocky Mountains during the winter months. This redeveloped low-pressure center began moving slowly east across the southern Great Plains and the lower Mississippi River Valley during the next two days. By 16 February it was located west of the southern end of the Appalachian Mountains and had started to decay once more. However, at the same time low pressure was developing just off the coast of North Carolina, along the *frontal system* of the original low-pressure center. The coastal low developed to become the main system, and moved north and east just offshore, slowly enough to allow large amounts of snow to fall. All-time record snowfalls were reported in the cities of Baltimore, Maryland (71.6 cm) and Boston, Massachusetts (69.9 cm).

What were the atmospheric processes responsible for the formation and evolution of this storm as it passed over the United States? Why were such diverse severe weather events as heavy snow and tornadoes reported with this one system? The study of atmospheric dynamics provides us with the tools to answer these questions, and more broadly helps us understand the processes that shape the circulation of the atmosphere. This book will use this storm as a guide in our exploration of the processes and forces that shape the circulation of the atmosphere. Weather maps that depict the evolution of the storm are provided on the accompanying CD-ROM.

1.2 Describing the atmosphere

We can consider the air that surrounds the Earth to be made up of 'columns', rising vertically from each location. As we move up through the column, the properties of the air change – its temperature, moisture, cloudiness, chemical constituents, and density all vary. The portion of the atmosphere of most interest in this book is the *troposphere*, the zone between the surface and around 10 km altitude. This is the part of the column in which 'weather' happens, and certainly where we experience it.

Before we begin our exploration of the effects of forces upon the atmosphere, the *dynamics*, we need to define some of its basic properties. These properties are essential for describing and understanding the atmosphere and can be thought of as the basic vocabulary of atmospheric dynamics.

The most familiar properties are wind speed, temperature, and density. As with all physical quantities, these are quantified using SI (Système International) or metric units. The metric system is founded upon a set of base units, of which we will use five: meter (m), kilogram (kg), second (s), kelvin (K), and mole (mol). Other units, such as for acceleration or force for example, can be derived from these basic units. There are two important quantities whose units are not in everyday use, and these require some further explanation: temperature and pressure.

1.2.1 Atmospheric temperature

Most commonly used around the world today is the Celsius scale (°C) for temperature, which was devised by Swedish astronomer Anders Celsius in 1742. The scale designates 0°C as the freezing point of water, and 100°C as the boiling point of water at sea level. The scale is also known as the 'centigrade' scale because it divides the interval between freezing and boiling into 100 equal units. However, this was not the scale adopted in the Système International.

In fact, in the early eighteenth century, there were many competing temperature scales. A second scale that has survived to the present day in the United States, and unofficially in many other English-speaking countries, is the Fahrenheit scale (°F). This temperature scale is named after the German physicist who suggested it in 1724.

The two references Fahrenheit chose were 0°F for the lowest temperature he could achieve and 100°F for body temperature. However, there is some dispute as to how he arrived at these reference points, and some time later the scale was recalibrated. On the present-day Fahrenheit scale, the freezing point of pure water is 32°F and the boiling point (at sea level) is 212°F.

The Kelvin temperature scale that was adopted in the Système International was designed by British scientist William Thomson (later Lord Kelvin) based on a single reference point, *absolute zero*, the hypothetically lowest temperature possible. This temperature was derived from an extrapolation of a graph showing the relationship between volume and temperature of an ideal gas, which suggested that the gas volume would become zero at a temperature of -273.16°C. The scale is often called the *absolute temperature*. Each unit on the Kelvin scale is the same magnitude as one degree on the Celsius scale, and hence the freezing point of water is around 273 K and the boiling point is around 373 K.

As we will see in Chapter 5, the physical equations we derive for application to the atmosphere assume SI units for all quantities. However, because the measurement and study of the weather has an important role across society, SI units are not always used in the reporting of the atmospheric state. This is particularly true for temperature, but also for wind speed and atmospheric pressure. Hence, it is important to be familiar with all of the units in general use, and to be able to convert between them (see Appendix B).

1.2.2 Atmospheric pressure

Of all of the properties of the atmosphere, atmospheric pressure is one of the most important. Horizontal and vertical variations in pressure give rise to the atmospheric motions which are the focus of the study of atmospheric dynamics.

Consider some air in a container. The pressure of the air on the walls of the container derives from the momentum of individual molecules as they impact the walls in their random molecular motion. If we add more molecules to the container, and that container happens to be a balloon, the difference in pressure between the interior and the exterior of the balloon will cause it to expand until a new equilibrium is reached (see Figure 1.1). Similarly, one infinitesimal volume parcel of air exerts pressure on its neighbor, and vice versa, and this force is always perpendicular to the interface between the parcels.

Hence, the pressure depends not only on the force imparted by the molecules, but also upon the area over which the force is acting. Thus, the pressure is expressed as a force per unit area, or $N\,m^{-2}$, which has the SI derived unit of the *pascal*, or Pa. Since 1 Pa is a small pressure in comparison to pressures commonly observed in the atmosphere, the unit hectopascal (symbol hPa) is more widely used. The unit millibar (symbol mb) is also in common use, particularly on weather maps:

$$1\,hPa = 100\,Pa = 1\,mb$$

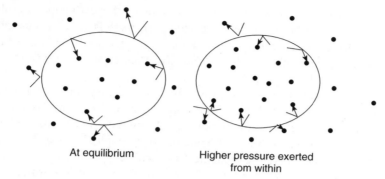

Figure 1.1 The balloon on the left has as many molecules exerting pressure from within as from without and hence is at equilibrium. The balloon on the right contains more molecules in the same volume and so (given the same temperature) is exerting more pressure from within. This balloon's walls will expand under the action of this force

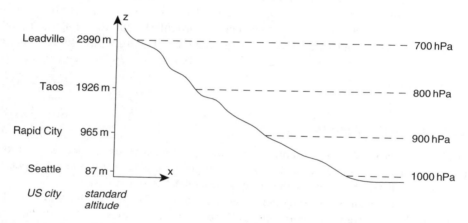

Figure 1.2 The weight of air in the column above any altitude applies a pressure on the surface below it. The axis on the left shows the standard altitude for the pressure level, and a representative US city at that elevation, for comparison

The *atmospheric pressure* at a point on the surface of the Earth is the pressure caused by the weight of air above that point (this is discussed in greater detail in Section 4.3.4). This implies that the atmospheric pressure must drop as one moves to higher altitudes (Figure 1.2) since there is less air above to exert its weight on a given point. This was first noted in 1648 by Blaise Pascal, after whom the unit of pressure is named, who made measurements of atmospheric pressure from the base to the top of the 1465 m Puy de Dôme mountain in France.

Because of this strong variation with altitude, pressure reports from meteorological stations are generally normalized to a common altitude, usually the mean sea level. This allows us to distinguish the smaller, but more important, horizontal variations

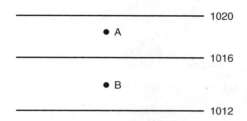

Figure 1.3 Idealized example of isobars, labeled in hPa, on a surface weather map

in surface pressure which are the ultimate cause of most motions observed in the atmosphere. The correction from surface pressure to mean sea level pressure (MSLP) is performed using the hypsometric equation, which is discussed in Section 4.3.4.

A line drawn on a weather map connecting points of equal pressure is called an *isobar*. Isobars are generated from reports of the pressure at various locations. Figure 1.3 depicts three isobars that map the observed MSLP in a particular region. At every point along the middle isobar, the pressure is taken to be 1016 hPa, while at every point along the bottom isobar, the pressure is 1012 hPa. Point A and all other points above the 1016 mb isobar have a higher pressure than 1016 hPa and points below that isobar have a lower pressure. At point B, and any other point lying in between these two isobars, the pressure must be between 1012 and 1016 hPa. Surface pressure reports are available as often as every hour in some countries, and usually every 3 or 6 hours. Isobar maps are updated with every new set of observations.

1.2.3 Station reports

For all surface weather maps around the world, the weather observations at a given location are reported using a common *station model*, an example of which is shown in Figure 1.4. The basic station model provides information on the temperature, dew point temperature, sea level pressure, wind speed and direction, cloud cover, and current significant weather. The *dew point temperature* (T_d) is used by operational meteorologists to indicate the amount of moisture contained in the atmosphere. It is defined as the temperature to which a small volume (or parcel) of air must be cooled at constant pressure in order for that air parcel to become saturated. If air near the surface of the Earth is cooled to its dew point temperature, dew will form on the surface. Higher values of dew point temperature indicate air that contains more water vapor than air with a lower dew point temperature.

The *wind speed symbols* are additive, and can be used to represent any wind speed, often reported in units of *knots* ($1 \, \text{kt} = 0.51 \, \text{m s}^{-1}$). A short wind barb indicates a wind speed of 5 kts, a long barb indicates a wind speed of 10 kts, and a pennant indicates a wind speed of 50 kts. It should be noted that the actual wind may be within ± 2 kts of the plotted wind speed. For example, if a station model is plotted with one

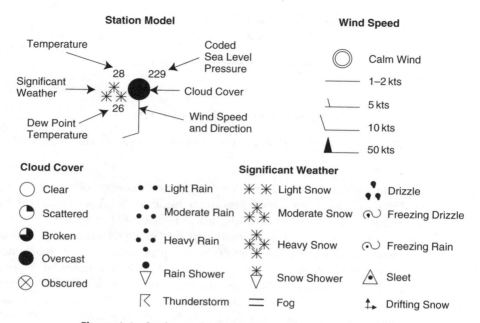

Figure 1.4 Station model for surface weather map conventions

short and one long wind barb the wind speed would be noted as 15 kts, but the actual wind speed could be anywhere from 13 to 17 kts. Similarly, a station model plotted with one pennant and two long barbs would have a nominal wind speed of 70 kts, but the actual wind speed could be anywhere from 68 to 72 kts. The wind direction is given by the angle of the line that anchors the barbs; the direction in which the line points is the direction the wind is coming *from*.

Sea level pressure is plotted in a coded format. To decode the sea level pressure report, either a 10 or 9 must be placed in front of the three-digit value, and then a decimal point added between the last two digits. If the coded sea level pressure is more than 500, a 9 is placed in front of the coded value. Otherwise, a 10 is placed in front of the coded value. The decoded sea level pressure is given in units of hPa, and using this coding system, it must range from 950.0 to 1050.0 hPa. In intense extra-tropical cyclones or tropical storms the sea level pressure may be less than 950 hPa. Similarly, strong high-pressure systems (anticyclones) may occasionally have sea level pressure values greater than 1050 hPa. For these situations care must be taken when decoding sea level pressure values given on weather maps.

Example What is the weather being reported in the station model shown in Figure 1.4?

First, we can see that the temperature is 28°F (−2°C) and the dew point temperature is 26°F (−3°C), corresponding to a relative humidity of around 92%. Why do we assume this is a US model? We can do so because the significant weather icon tells us that moderate snow is falling, and this is highly unlikely at 28 °C (82 °F)!

We also see that the wind is blowing from the south at around 10 kts, a brisk southerly. The skies are overcast, as we would expect while it is snowing. The coded sea level pressure is 229; this is less than 500 and so we place a 10 at the start to get 10229, and then add a decimal point between the last two digits to give us 1022.9 hPa. In total, a chilly and gray day is being reported.

As well as conforming to a standard set of observations, stations around the world also conform to a standard description of the time the observation was taken: *Universal Time (UTC)*. This time, also known as *Greenwich Mean Time (GMT)*, or simply Z, is the time at Greenwich in England, and is expressed using a 24 hour clock rather than a.m and p.m. UTC is used regardless of the local time zone of the observation.

1.3 Air masses and fronts

In the early and middle parts of the last century, weather forecasters began to understand the systematic nature underlying the apparent chaos of the weather, and to use it to advantage in weather forecasting. The foundation of this understanding was the dual concepts of air masses and fronts. Air masses, and the fronts between them, are defined by their thermodynamic properties; that is, their temperature, density, and pressure, and the amount of moisture they contain. The critical idea that developed was that weather conditions are not randomly distributed over the globe. Rather, variations in weather elements, such as temperature, in the extra-tropics tend to be concentrated in narrow bands called *fronts*. Between these bands, weather elements change very gradually, and these broad, nearly uniform regions of the lower atmosphere are called *air masses*.

1.3.1 Air masses

The best conditions for the formation of air masses are large areas where air can be in relatively constant conditions long enough to take on quite uniform characteristics. The warm, moist tropical oceans and the cold, dry polar land masses are excellent source regions. An air mass can be relatively shallow, 1–2 km deep, or may be as deep as the entire troposphere, and is often not of uniform depth across its entire horizontal extent.

A commonly used *classification of air masses*, especially in the Northern Hemisphere, is that of Tor Bergeron of the Norwegian School (see Section 1.4), who in 1928 denoted four primary air masses – continental polar (cP), continental tropical (cT), maritime polar (mP), and maritime tropical (mT). Often, additional air masses such as the Arctic (A), Antarctic (AA), and equatorial (E) are added to this list. Figure 1.5 shows the source region and typical locations impacted by these different air masses. These air masses are identified either subjectively or objectively using a range of criteria,

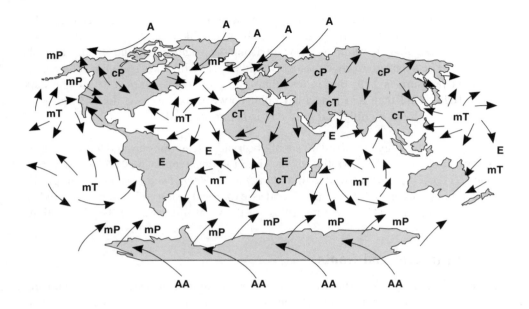

Figure 1.5 Typical distribution of air masses around the globe

but are typically identified based on near surface values of temperature and humidity. A qualitative description of the typical temperature and humidity of the different air mass types is also given in Figure 1.5. Since air masses are three-dimensional features of the atmosphere they may also be identified using atmospheric observations from balloons with instruments attached, called *radiosondes*, or by aircraft.

At any one time there are at least several dozen distinct air masses globally. Most cover thousands of square kilometers of surface area. Some have just formed, some are in the process of being modified, and some are essentially stationary.

Once an air mass moves out of its source region, it begins to be modified as it encounters surface conditions different from those found in the source region. From the point of view of atmospheric dynamics, air masses are generally of interest primarily because of the interfaces between them: the fronts.

1.3.2 Fronts

Fronts are the boundaries that separate different air masses, and are defined by thermodynamic differences across the boundary, and the direction of movement of

AIR MASSES AND FRONTS

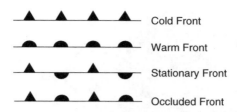

Figure 1.6 Symbols used to represent fronts on weather maps. The side on which the symbols are drawn is the direction in which the front is traveling, except for the stationary front, which is not moving

the boundary. Meteorologists define four basic *types of fronts*, although other types of fronts can also be identified. These four basic types of fronts are referred to as cold, warm, stationary, and occluded fronts. The symbols used to represent these fronts on weather maps are shown in Figure 1.6.

Typically fronts separate warm and cold air masses. If the cold air mass is advancing and the warm air mass is retreating the boundary is called a *cold front*. If the opposite occurs, with warm air advancing and cold air retreating, the boundary is called a *warm front*. Sometimes the boundary between the two air masses is nearly stationary and this type of front is called a *stationary front*. An *occluded front* separates air masses that do not have as large a temperature contrast as is found for cold or warm fronts, and typically separates cold and cool air masses. The processes that lead to the formation of an occluded front are discussed below, and are important in the life cycle of extra-tropical cyclones.

Even though fronts are most commonly seen only on surface weather maps it is important to remember that fronts are the boundaries that separate three-dimensional volumes of the atmosphere known as air masses, and can be identified both at the surface and aloft. An example of a warm, cold, and stationary front associated with the February 2003 cyclone (Section 1.1) is shown in Figure 1.7. Note how the warm front and the cold front meet at the low-pressure center.

This is a *surface weather map* of the type frequently used in the United States, showing isobars, fronts, and station reports, which indicate temperature and dew point temperature in °F, wind speed observations in knots, coded sea level pressure, and cloud cover. Weather maps used in other parts of the world typically report the temperature and dew point temperature in °C, but are otherwise equivalent. You can find examples of these types of weather maps on the companion CD-ROM.

Example Identify the warm and the cold air masses in the map shown in Figure 1.7.

Consider the station observation just to the south-east of the center of the low. The wind is 5 kts from the SSE, the weather is clear, and the temperature is 60°F (16°C). The surface pressure is 1001.1 hPa. Just to the north of this station, there is a station on the other side of the warm front. Here, the weather is overcast and foggy, and the wind is from the NE at 5 kts. The temperature is 44°F (7°C). The *dew point*

12 ANATOMY OF A CYCLONE

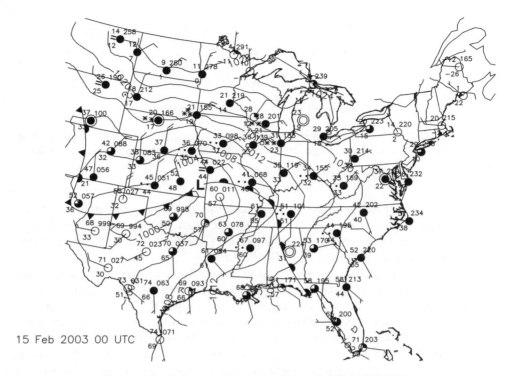

15 Feb 2003 00 UTC

Figure 1.7 Surface weather map valid at 00 UTC 15 Feb 2003

depression, the temperature minus dew point temperature, is zero, meaning that the air is saturated. It is clear that south of the low, between the cold front and the warm front, is the warm air mass. North, and to either side of the fronts, is the cold air mass.

1.3.3 Upper level weather maps

Meteorologists around the world also use standard format maps to depict atmospheric features higher in the troposphere. By convention, these upper level maps are plotted on constant pressure surfaces rather than at specific altitudes, and hence are called *isobaric* (iso=constant; baric=pressure) maps. A convenient upper level map to look at when examining an extra-tropical cyclone is the 500 hPa map, which represents atmospheric conditions at an altitude of approximately 5.5 km. This is roughly in the middle of the troposphere, and the winds at this level are often thought to 'steer' weather disturbances.

An example of a 500 hPa map, for the same day and time as the map in Figure 1.7 is shown in Figure 1.8. The station reports plotted on this map replace the pressure code with the height (that is, the altitude) of the 500 hPa surface. For the 500 hPa constant pressure level the coded height is given in decameters (tens of meters), but the coding method varies from one constant pressure level to another. On upper level

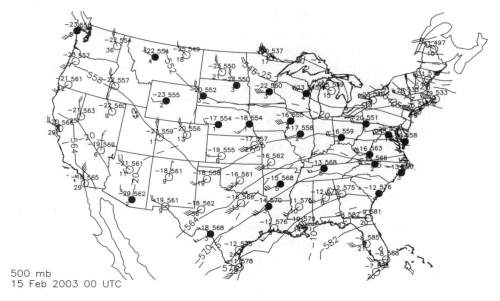

Figure 1.8 The 500 hPa weather map valid at 00 UTC 15 Feb 2003. Height contours (decameters) are shown as solid lines, and temperature contours (°C) are shown as dashed lines

maps, the temperature is always plotted with units of °C throughout the world, and as for surface weather maps is plotted in the upper left corner of the station model. The dew point depression, rather than the dew point temperature, is plotted in the lower left corner of the station model. The central circle of the station model is filled if the dew point depression is less than 5°C, since a small dew point depression is indicative of areas that may be cloudy. Otherwise, the circle is left unfilled.

From Figure 1.8 we see that the height of the 500 hPa surface varies, with lower heights located to the north and higher heights found to the south. There are a number of ripples, or waves, superimposed on this general decrease of height from south to north. The height contours on the 500 hPa map can be thought of as being similar to the elevation contours found on topographic maps of the Earth's surface. Locations where low heights are found are referred to as *troughs* since they represent a dip in the elevation of the 500 hPa surface. *Ridges* are regions where high heights are found on the 500 hPa map. Small ripples in the 500 hPa heights are known as *short waves* while larger ripples, that extend across a whole continent for example, are known as *long waves* or *planetary waves*. In this example, a ridge is present over the west coast of the United States, while a long-wave trough is found over the center of the country.

The height of the 500 hPa surface is related to several properties of the atmospheric column. For example, as the temperature of the column decreases, the height of the 500 hPa surface would also tend to decrease (see Section 4.3.4 for more details). That is, troughs are cold, and ridges are warm, in general. This leads to lower 500 hPa heights being found near the poles and higher heights being found near the tropics.

The wind at this level generally flows roughly parallel to the height contours. Higher wind speeds are found in areas where the height contours are closely spaced and weaker wind speeds are found in areas where the height contours are more widely spaced. We will see why this is the case in Section 5.5. In the Northern Hemisphere, the winds blow with lower heights to the left of the wind direction and higher heights to the right of the wind direction. In the Southern Hemisphere, the winds blow with lower heights to the right of the wind direction. Because the 500 hPa surface slopes downward from the equator toward the poles, this means that a large component of the wind is blowing from the west almost all the time in both hemispheres. These winds are aptly called the *mid-latitude westerlies*.

In addition to the height contours on the 500 hPa map shown in Figure 1.8, contours of constant temperature, or *isotherms*, are shown. As would be expected, lower temperatures are found closer to the poles while higher temperatures are found closer to the tropics. An atmosphere in which the temperature varies across a pressure surface is referred to as being *baroclinic*. In a baroclinic atmosphere the density of air depends upon both temperature and pressure (see Section 3.2 and Equation (3.1) for more details). An atmospheric state in which temperature does not vary on a constant pressure surface is referred to as *barotropic*, and in this case density is only a function of pressure. It is a situation that occurs in the atmosphere only rarely and approximately.

1.4 The structure of a typical extra-tropical cyclone

The cyclones of middle and high latitudes are called extra-tropical cyclones. They differ from cyclones in the tropics in a number of ways, but most prominent is the fact that extra-tropical cyclones contain frontal systems while tropical cyclones do not. Consequently, extra-tropical cyclones are also known as *frontal cyclones*. In this chapter we will discuss only cyclones from the middle latitudes, reserving discussion of tropical cyclones for Chapter 12 and polar cyclones for Chapter 14.

No two cyclones are identical, but an idealized model was developed during World War I in Norway which embodies many of the important features of a frontal cyclone. See Bjerknes and Solberg (1922) in the Bibliography for the original description, which is drawn upon in this discussion. This model is known as the Norwegian model, or the Bjerknes model. The model was not widely accepted in recent decades, until research since the 1990s confirmed that the model remains a useful tool to describe the weather associated with extra-tropical cyclones. In this section, and throughout this book, we will use this model to describe the extra-tropical cyclone that traversed the United States from 11 to 19 February 2003. However, as with all conceptual models, the Norwegian model does not always apply.

Frontal cyclones are large traveling atmospheric vortices (rotating air), up to 2000 kilometers in diameter, with centers of low atmospheric pressure. An intense cyclone in middle to high latitudes may have a surface pressure as low as 970 hPa, compared to an average sea level pressure of 1013 hPa.

Frontal cyclones are the dominant weather event of the Earth's middle latitudes, and are the environment in which smaller, more intense circulations are often embedded. In the middle latitudes, they are the result of the dynamic interaction of warm tropical and cold polar air masses. The boundary between these two air masses is called the *polar front*. This larger frontal zone is distinct from the individual cold and warm fronts associated with a particular cyclone. We will now look at the idealized life cycle of an extra-tropical cyclone, as portrayed in the Norwegian model. We will need to consider not just the surface, but the middle and upper regions of the troposphere, since cyclones are three-dimensional features of the atmosphere.

1.4.1 Stages in the life cycle of an extra-tropical cyclone

As discussed in Section 1.2.2, the pressure at any location on the surface of the Earth is determined by the mass of air in the atmospheric column above. For surface pressure to decrease, air must be removed from the atmospheric column. Therefore the formation, intensification, and decay of the surface low-pressure centers that define cyclones are intimately tied to processes that add atmospheric mass (*convergence*) or remove atmospheric mass (*divergence*) from a column of the atmosphere.

According to the Norwegian school, extra-tropical cyclones form along the polar front. In this idealized model, the polar front initially runs west to east in a more or less straight line until a small perturbation in the form of a wave in the upper levels of the troposphere disturbs it (Figure 1.9). Divergence occurs in the vicinity of this upper level short-wave trough (Figure 1.9a). This divergence promotes the initial formation of the surface cyclone.

In the Northern Hemisphere, the circulation around this area of low pressure is counterclockwise and results in cold air moving southward on the west side of the surface cyclone, with the leading edge of the cold air marked by a cold front. East of the surface cyclone warm air is moving northward, with the leading edge of the warm air marked by a warm front (Figure 1.9b). The result is that the small wave along the polar front amplifies, and the cyclone is referred to as being an *open wave cyclone*.

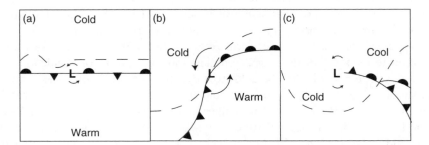

Figure 1.9 Idealized life cycle of a mid-latitude cyclone. L marks the position of the surface low-pressure center and arrows indicate the surface air circulation. Fronts are shown using standard symbols. A representative 500 hPa height contour is shown as a dashed line

The wedge of air between the advancing cold and warm fronts (south and east of the surface cyclone in Figure 1.9b) is known as the *warm sector* of the cyclone. We discovered this sector in the example at the end of Section 1.3.2. The air mass in this sector is characterized by temperature and humidity values that are larger than behind the cold front or ahead of the advancing warm front.

At upper levels the heights of the constant pressure surfaces begin to respond to the movement of the cold and warm air masses. The height of the constant pressure surface begins to decrease in regions where cold air is advancing, forming a deeper upper level trough to the west of the surface cyclone. Conversely the height of the constant pressure surface begins to rise in regions where warm air is advancing, resulting in the formation of an upper level ridge ahead of the surface cyclone. The cyclone shows a pronounced tilt toward the west with increasing height at this time, with the upper level trough west of the surface cyclone position. This is the mature stage of the cyclone life cycle, and may last up to a few days. Figure 1.10 shows an idealized representation of a mature extra-tropical cyclone, and can be compared with the mid-latitude cyclone shown in Figure 1.7. The balance between upper level divergence and lower level convergence will determine how quickly the cyclone intensifies. During this time the cyclone tends to move in a direction that is parallel to the 500 hPa winds, and so the cyclone in Figure 1.9b will move north-eastward.

An animation of the three-dimensional circulation around a mature extra-tropical cyclone is provided on the companion CD-ROM. In this animation both the horizontal and vertical motion associated with the cyclone are shown, and the tilt of the low pressure towards the west with increasing height can be seen. Rising motion occurs near the center of the surface cyclone and ahead of the warm front, leading to the formation of clouds and precipitation. Clouds tend to be low along and just ahead of the warm front. As one moves further ahead of the warm front the clouds become higher and thinner and the precipitation decreases in intensity and eventually stops. Along the cold front there is a narrow band of rising motion associated with the cold air undercutting the air in the warm sector of the cyclone (not shown on the animation), and a narrow band of clouds and precipitation can be expected along the cold front.

The cold and warm fronts will continue to advance in response to the circulation around the surface low-pressure center as it deepens. The cold front often advances more quickly than the warm front, with the cold air undercutting the warm air in the warm sector of the storm. As the warm front advances, the warm air rises up over the cold air to the north of the warm front, and thus the warm front tends to advance more slowly. The result is that the cold front eventually catches up to the warm front at the surface (Figure 1.9c). Where the cold front has caught up to the warm front, the warm air is forced aloft and is no longer found at the surface. The boundary that was the leading edge of the cold front now separates two regions of cold air at the surface, and is called an *occluded front*. The cyclone is now said to be occluded and is often referred to as a *cold core* cyclone. In response to the change in the surface air temperature distribution the 500 hPa trough is now centered over the surface cyclone. The cyclone has now reached the end of its life cycle, and will slowly decay.

THE STRUCTURE OF A TYPICAL EXTRA-TROPICAL CYCLONE

Vertical cross-section from A to B

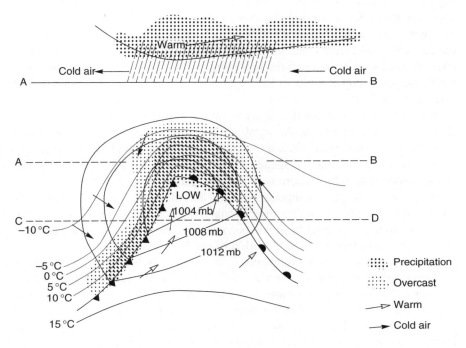

Vertical cross-section from C to D

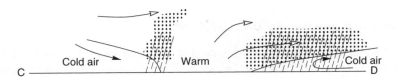

Figure 1.10 An idealized representation of a mature extra-tropical cyclone (adapted from the work of Bjerknes and Solberg 1922)

As with all models, this idealized description of the typical life cycle of an extra-tropical cyclone does not always match what is observed, and has been modified over time. Additionally, not all cyclones go through the stages illustrated in this idealized model. Some cyclones remain as small frontal waves which *propagate* (travel) rapidly eastward, while other cyclones grow rapidly but then decay before attaining the cold core occlusion stage. In some cases the cyclones become massive storms which are accompanied by changes in the entire long-wave pattern aloft. Other conceptual models of mid-latitude cyclones are in use by meteorologists and atmospheric scientists.

1.4.2 A mature cyclone – 00 UTC 15 Feb 2003

Figures 1.7 and 1.8 show the surface and 500 hPa weather maps at 00 UTC 15 Feb 2003 for this cyclone during the mature stage of its life cycle. The surface low-pressure center is located in south-eastern Kansas, with a warm front extending to the east and a cold front extending to the south-west. The warm sector of the storm is characterized by air temperatures in the upper 60s and lower 70s °F (around 20°C), with dew point temperatures in the upper 50s and lower 60s °F (around 15°C). The surface winds in the warm sector are primarily from the south or south-east.

North of the warm front the air temperature and dew point temperature are lower than in the warm sector, and the winds are from the east or north-east. Overcast skies extend from the warm front north to central Minnesota and North Dakota, associated with rising motion along and ahead of the warm front. Light rain is falling just north of the warm front, with light to moderate snow falling further north.

The air and dew point temperatures drop abruptly across the cold front, while the winds shift to the north and become strong and gusty. Along the eastern edge of the Rocky Mountains the cold front has become stationary, as is typical when a shallow cold front encounters the high elevation of the Rocky Mountains.

At the 500 hPa level a broad long-wave trough is located west of the surface cyclone position, generating south-westerly flow aloft above the surface cyclone. The positions of the surface lower and upper level troughs indicate that the cyclone is in a favorable position for further development at this time, but the cyclone maintains a nearly constant intensity over the next 48 hours, indicating that divergence at upper levels is nearly balanced by convergence of air near the surface.

1.4.3 An occluded cyclone – 00 UTC 17 Feb 2003

By 00 UTC 17 Feb 2003 the surface cyclone that had been located in south-eastern Kansas is now located on the Kentucky/Tennessee border just west of the Appalachian Mountains (Figure 1.11). A weak occluded front extends southward from the surface low-pressure center and the low is surrounded by cool air at the surface. This low has become occluded, and will slowly weaken and drift north-eastward over the next two days.

Two new surface low-pressure centers are beginning to form, with one along the Georgia/Alabama border and the other just offshore from North Carolina. Both of these lows have formed along the frontal boundaries associated with the original low-pressure center. A broad area of cloudiness and precipitation is associated with the occluded low and the frontal boundaries. North of the surface low pressure offshore of North Carolina winds are from the east, bringing moisture from the Atlantic Ocean, resulting in light snow.

At 500 hPa, a long-wave trough is located nearly overhead of the occluded surface low (Figure 1.12), as we would expect from the Norwegian cyclone model. The surface low off of the North Carolina coast is in a favorable location for further development, being located to the east of the 500 hPa trough. Further, the contrast

THE STRUCTURE OF A TYPICAL EXTRA-TROPICAL CYCLONE

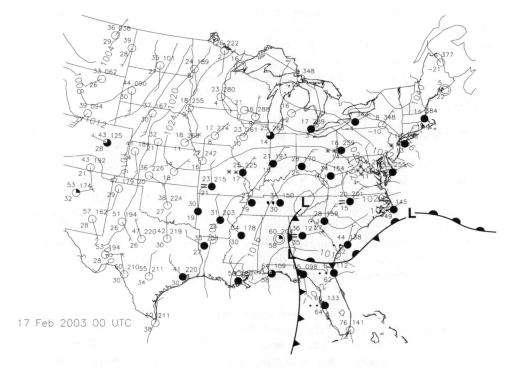

Figure 1.11 Surface weather map valid at 00 UTC 17 Feb 2003

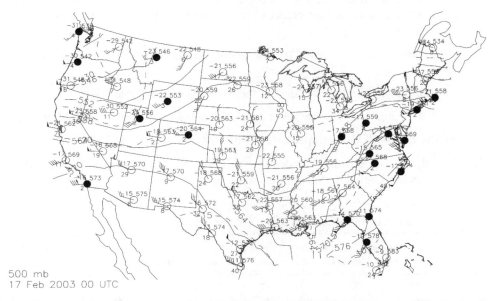

Figure 1.12 The 500 hPa weather map valid at 00 UTC 17 Feb 2003. Height contours (decameters) are shown as solid lines, and temperature contours (°C) are shown as dashed lines

between the relatively warm water of the Atlantic Ocean and the cold continental air will serve to increase the temperature contrast across the fronts associated with this new surface cyclone. In response to these factors that favor intensification of the surface low pressure, this low does deepen over the next two days bringing heavy snow to the mid-Atlantic and the north-eastern United States.

Review questions

1.1 In preparing a surface weather map over the Antarctic continent, station models for all available surface observations must be constructed. At 21 UTC 25 Sep 2005 the surface weather observation at McMurdo Station (77.9°S, 166.7°E) indicated a temperature of −36°F, a dew point temperature of −45°F, a sea level pressure of 988.4 hPa, and calm winds. At the same time the surface weather observation from Dumont D'Urville (66.7°S, 140.0°E) indicated a temperature of −11°F, a dew point temperature of −19°F, a sea level pressure of 982.9 hPa, and a wind from the south-east at 10 kts. Draw the station models for these two sets of observations using the surface station model plotting conventions for US surface weather maps. Draw the same station models using international plotting conventions.

1.2 Examine the map of surface observations shown in Figure 1.13 on the CD-ROM. One way of identifying air masses is to look for regions of warmer and colder temperatures. Use only the temperature field to draw a line on the map that marks the boundary between the continental polar air mass and the maritime tropical air mass.

Now use the additional information available in the surface observations to redefine the air masses using a second copy of the map. What information did you use? Are there any differences between your first analysis and your second analysis? If so, explain why.

1.3 What type of air mass would be responsible for the weather conditions listed as follows?

(a) hot, muggy summer weather in the Midwest of the United States;

(b) refreshing, cool, dry breezes after a long summer hot spell in central Europe;

(c) drought with high temperatures in the Sahel region of Africa.

1.4 Consider the wintertime climates of Rochester, New York and Rochester, Minnesota in the United States (Table 1.1).

Table 1.1 The wintertime climates of Rochester, MN and Rochester, NY

	January average maximum (°C)	January average minimum (°C)	Annual snowfall (m)	Latitude	Elevation (m)
Rochester, MN	−7	−17	1.17	44°01′18″N	1317
Rochester, NY	−3	−9	2.29	43°09′17″N	50

Find the two cities on a map. What are the causes of the differences between the winters experienced by people in these two cities?

1.5 Consult the weather map shown in Figure 1.14 on the CD-ROM – it shows an occluded low-pressure system in the Greenland Sea. An ocean report in the cold air mass, west of the occluded front, shows overcast skies and drizzle. Identify this station report and decode the temperature, dew point temperature, sea level pressure, wind speed and wind direction.

1.6 Using the surface weather map for 00 UTC 16 Feb 2003 on the CD-ROM decode the surface station models for:

(a) Dallas, Texas

(b) Atlanta, Georgia

(c) Dyersberg, Tennessee.

1.7 Print a copy of the surface weather map for 00 UTC 16 Feb 2003 from the CD-ROM.

(a) Draw isotherms every 10°F on this map.

(b) Indicate the position of the low-pressure center by marking a letter 'L' on your map.

(c) Draw cold and warm fronts using all of the data displayed on this map.

(d) Where are the weather stations used in question 1.6 located relative to the fronts you identified in question 1.7(c)? Are the weather observations you decoded for question 1.6 consistent with your expectations of the weather associated with the passage of the fronts?

1.8 (a) Why is the dew point always lower than the actual temperature?

(b) Why does a small dew point depression indicate areas that are likely to be cloudy?

1.9 Decode the upper air station models at Denver, Colorado and Atlanta, Georgia plotted on the 500 hPa map for 00 UTC 16 Feb 2003 provided on the CD-ROM.

1.10 Identify the position of the troughs and ridges on the 500 hPa map for 00 UTC 16 Feb 2003 on the CD-ROM.

(a) Where is the surface low-pressure center identified in question 1.7(b) located relative to the troughs and ridges at 500 hPa?

(b) Is this location favorable for the low-pressure system to intensify? Why or why not?

(c) How has the intensity of the low-pressure system changed by 12 UTC 16 Feb? Is this change in intensity consistent with your answer to question 1.10(b)?

2 Mathematical methods in fluid dynamics

This chapter presents several useful mathematical definitions, particularly associated with vectors and functions in multiple dimensions. Many of these definitions should be familiar to the student, and hence will be reviewed briefly and without detailed proofs. The important concepts of Eulerian and Lagrangian frameworks, and the associated process of advection, will be discussed in greater detail.

2.1 Scalars and vectors

A scalar is any quantity which can be fully specified by a single number, its magnitude. Examples include time, temperature, wind speed, and surface pressure.

Some quantities in nature require both a magnitude and a direction to be fully specified. Wind velocity is such a quantity, which has a magnitude (the wind speed) and a direction. Together this description forms a vector quantity. There are two typical notations for a vector:

- Graphically: A vector pointing from location A to location B is written as the vector **AB**. The vector has a magnitude given by the distance between A and B, and a direction given by the direction of the line in a given coordinate system.

- In terms of components: The vector $\vec{u} = u_x \vec{i} + u_y \vec{j}$ is a vector with two components, one in the x or $\vec{i}$ unit vector direction with magnitude u_x, and one in the y or $\vec{j}$ unit vector direction with magnitude u_y, in a given coordinate system. The magnitude, denoted by $|\vec{u}|$, is given by $\sqrt{u_x^2 + u_y^2}$, and the direction of the vector is $\tan^{-1}(u_y/u_x)$.

2.2 The algebra of vectors

In any mathematical theory, we should begin by defining what is meant by addition, subtraction, and multiplication of our new quantities, vectors. We can define these

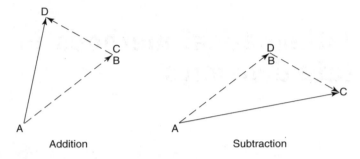

Figure 2.1 The addition and subtraction of vectors **AB** and **CD**

operations graphically or computationally, and it is useful to consider both of these approaches.

2.2.1 Addition and subtraction

To add the vector **AB** to vector **CD**, we can place **CD** so that location C, the starting point of **CD**, falls on location B, the end point of **AB** (Figure 2.1). Then the sum of **AB** and **CD** is the vector **AD**. Vector addition is both commutative and associative. This graphical definition of a vector sum can be very useful when we are considering how various forces combine to produce the motion of a parcel of air. However, sometimes we need to compute a sum, or we need to perform an addition in more than two dimensions. In such cases, it is more straightforward to write, for example,

$$\vec{u} = u_x \vec{i} + u_y \vec{j} + u_z \vec{k} \quad \text{and} \quad \vec{v} = v_x \vec{i} + v_y \vec{j} + v_z \vec{k}$$
$$\Rightarrow \vec{u} + \vec{v} = (u_x + v_x)\vec{i} + (u_y + v_y)\vec{j} + (u_z + v_z)\vec{k} \tag{2.1}$$

for any two vectors with three components in the $\vec{i}$, $\vec{j}$, and $\vec{k}$ unit vector or x, y, z axis directions.

Conversely, to subtract one vector, **CD**, from another, **AB**, we simply add the negative of **CD**, which is **DC**, to **AB** (Figure 2.1). Computationally, this is simply

$$\vec{u} - \vec{v} = (u_x - v_x)\vec{i} + (u_y - v_y)\vec{j} + (u_z - v_z)\vec{k} \tag{2.2}$$

2.2.2 Multiplication by scalars

If **AB** is a vector and c is a scalar, then the product c**AB** is a vector whose length is c times the length of **AB**. This new vector has the direction **AB** if c is positive, and has the direction **BA** if c is negative (Figure 2.2). If c equals zero then c**AB** is the zero vector.

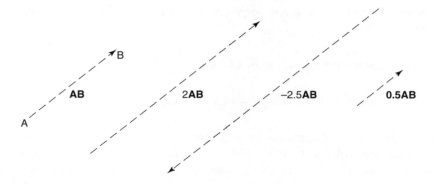

Figure 2.2 The multiplication of vector **AB** by scalars $c = 2$, -2.5, and 0.5

Writing this computationally in terms of components,

$$c\vec{u} = cu_x\vec{i} + cu_y\vec{j} + cu_z\vec{k} \tag{2.3}$$

2.2.3 Multiplication of two vectors

There are two different ways to formulate the multiplication of two vectors, both of which are relevant in atmospheric dynamics. The first is called the *scalar product* or dot product of two vectors, and this process results in a scalar. The second is called the *vector product* or cross product, the result being a vector. Again, we can define these functions both graphically and computationally. Let us first consider the scalar product, which is written

$$\vec{u} \bullet \vec{v} = u_x v_x + u_y v_y + u_z v_z = \left|\vec{u}\right|\left|\vec{v}\right| \cos\theta \tag{2.4}$$

where θ is the angle between the two vectors. Notice that if the two vectors are perpendicular, then $\cos\theta = 0$ and the scalar product between the vectors is also zero. Conversely, if the scalar product between two vectors is zero, and we know that the two vectors themselves are nonzero, then we know that they are perpendicular.

Example The scalar product of the two vectors

$$\vec{u} = 4\vec{i} - 2\vec{j} + 3\vec{k}$$
$$\vec{v} = 2\vec{i} - 2\vec{j} - 4\vec{k}$$

is

$$\vec{u} \bullet \vec{v} = (4 \times 2) + (-2 \times -2) + (3 \times -4)$$
$$= 8 + 4 - 12$$
$$= 0$$

and hence the two vectors are perpendicular. We can also see this graphically (Figure 2.3)

Computationally, the vector product is written

$$\vec{u} \times \vec{v} = (u_y v_z - u_z v_y)\vec{i} - (u_x v_z - u_z v_x)\vec{j} + (u_x v_y - u_y v_x)\vec{k} \qquad (2.5)$$

This definition does not appear to be very intuitive, but it results in a vector product which is perpendicular to both $\vec{u}$ and $\vec{v}$, and in fact forms a *right-handed set* with these two vectors (Figure 2.4). The magnitude of this vector product is $|\vec{u}||\vec{v}|\sin\theta$. This tells us that if $\vec{u}$ and $\vec{v}$ are parallel, their vector product will be zero. For the student who is familiar with the use of determinants, we can use an alternative form for this product

$$\vec{u} \times \vec{v} = \begin{vmatrix} \vec{i} & \vec{j} & \vec{k} \\ u_x & u_y & u_z \\ v_x & v_y & v_z \end{vmatrix} \qquad (2.6)$$

$$= (u_y v_z - u_z v_y)\vec{i} - (u_x v_z - u_z v_x)\vec{j} + (u_x v_y - u_y v_x)\vec{k}$$

which gives the same result.

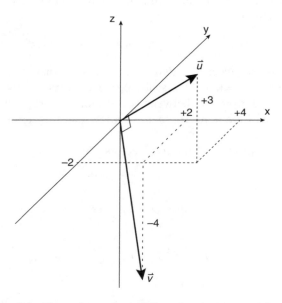

Figure 2.3 The scalar product of these two vectors must be zero

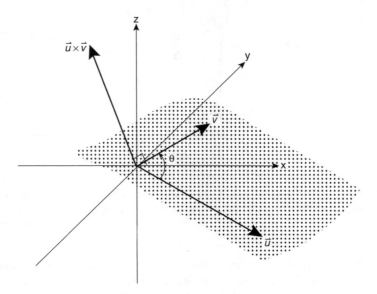

Figure 2.4 The vector product of two vectors is perpendicular to the plane containing the two vectors

2.3 Scalar and vector fields

A field is a quantity which is continuously defined over a given coordinate space. For example, in the map shown in Figure 1.7, a value for the surface pressure is known, or can be deduced, everywhere on the map.

In a scalar field, a scalar (or number) is assigned to every point in the space; the scalar may change from point to point and with time. In fact, the scalar is a function of the three coordinates of position and the one coordinate of time. Examples of scalar fields are temperature and pressure. Scalar fields can be shown physically on a map by lines joining points of equal value such as in lines of equal temperature (isotherms) or lines of equal pressure (isobars).

In a vector field, a vector is assigned to every point in the space; the vector may change from point to point and with time, so that the vector is a function of the three coordinates of position and one coordinate of time. Examples of vector fields include the wind and forces such as gravity and the pressure gradient force (Chapter 4). Vector fields are typically represented by arrows, where the length of each arrow indicates the magnitude. Meteorological maps will often represent vector fields with wind barbs as discussed in Chapter 1 (for example, Figure 1.8).

2.4 Coordinate systems on the Earth

A field of a scalar quantity, like temperature, does not change when the coordinate system is changed. Because of this property, scalar quantities are particularly easy to

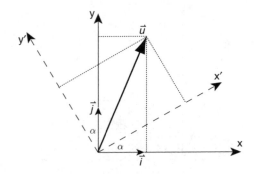

Figure 2.5 Effect of a coordinate system rotation on the components of a vector. The original coordinate system is denoted x–y and the rotated coordinate system is denoted x'–y'

manipulate mathematically. However, for a vector quantity, the components change under the rotation of the coordinate system. For example, consider the vector $\vec{u} = u_x \vec{i} + u_y \vec{j}$. Now, consider the vector in a rotated coordinate system as shown in Figure 2.5.

In the new coordinate system, the components of the vector become

$$\vec{u}' = (u_x \cos \alpha + u_y \sin \alpha)\vec{i}' + (u_y \cos \alpha - u_x \sin \alpha)\vec{j}' \qquad (2.7)$$

where α is the angle between the original and new coordinate systems as shown in Figure 2.5.

Since many of the quantities we will be interested in are vectors, it is important that we define our location and coordinate system precisely. In atmospheric dynamics, it is traditional to define the coordinate system relative to the Earth, even though the Earth is moving through space. In fact, the Earth is not just moving through space, it is accelerating, because its motion involves changes in direction even though the speed does not change on the time scales we are considering. This is termed a *non-inertial frame of reference*, and will have consequences for how we understand motion relative to the Earth. We will examine this in Chapter 4. The way in which we define this coordinate system is shown in Figure 2.6.

Thus, at any point on the Earth we define direction according to our departures from due east (the x direction), due north (the y direction), and the local vertical (the z direction).

2.5 Gradients of vectors

Consider a three-dimensional vector $\vec{u}(t)$ which represents the wind speed and direction in space at some time t. Then we can differentiate the vector function just

GRADIENTS OF VECTORS

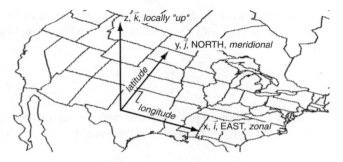

Figure 2.6 Coordinate system anchored on the Earth

as we differentiate a function in one-dimensional space. It is most straightforward to write the velocity vector in terms of its components:

$$\vec{u}(t) = f(t)\vec{i} + g(t)\vec{j} + h(t)\vec{k} \qquad (2.8)$$

Assuming that all of the component scalar functions are continuous and differentiable (and hence the velocity vector function itself is continuous and differentiable) we can write

$$\vec{u}'(t) = \frac{d\vec{u}(t)}{dt} = \frac{df}{dt}\vec{i} + \frac{dg}{dt}\vec{j} + \frac{dh}{dt}\vec{k} \qquad (2.9)$$

Example For the wind velocity vector defined by

$$\vec{u}(t) = (t^3 + 2t)\vec{i} + 4\vec{j} + \cos t\vec{k}$$

the acceleration vector, that is the derivative of the velocity vector with respect to time, is

$$\frac{d\vec{u}}{dt} = (3t^2 + 2)\vec{i} - \sin t\vec{k}$$

However, we are quite unlikely to come across a wind field that is uniform in space even as it changes in time. For example, as we move out from the shelter of a building on a windy day, the wind changes in both magnitude and direction. So, in general, we do need to consider wind fields (and other vector fields) that are functions of space (x, y, z) as well as time t; that is, three-dimensional vectors that are functions of four variables!

Just as in our earlier example, the derivative of a vector gives the rate of change of that vector. We can simplify our approach by only considering *partial derivatives*; that is, the derivative with respect to one independent variable while holding the other independent variables constant. Hence, for the x component of the wind vector function $f = f(x, y, z, t)$, to find only the variation of f with respect to t, we write the partial derivative as

$$\frac{\partial f}{\partial t}$$

So, taking the x, y, and z components of the wind velocity vector into consideration, the partial derivative of the entire wind vector with respect to time is written

$$\frac{\partial \vec{u}}{\partial t} = \frac{\partial f}{\partial t}\vec{i} + \frac{\partial g}{\partial t}\vec{j} + \frac{\partial h}{\partial t}\vec{k} \qquad (2.10)$$

Example For the wind velocity vector

$$\vec{u}(x, y, z, t) = (t^3 xz)\vec{i} + (4\sin xt + 5y^3)\vec{j} + (2xy^2)\vec{k}$$

The acceleration vector, that is the derivative of the velocity vector with respect to time, is

$$\frac{\partial \vec{u}}{\partial t} = (3t^2 xz)\vec{i} + (4x\cos xt)\vec{j}$$

Higher order partial derivatives are defined in the same way, so that for example

$$\frac{\partial^2 f}{\partial x^2} = \frac{\partial}{\partial x}\left(\frac{\partial f}{\partial x}\right) \quad \text{and} \quad \frac{\partial^2 f}{\partial x \partial t} = \frac{\partial}{\partial x}\left(\frac{\partial f}{\partial t}\right)$$

We can assume for our purposes that the order in which the partial derivatives are taken is generally unimportant.

It is useful at this point to consider graphically what partial derivatives in different spatial directions may mean. To simplify our picture, let us assume that we have a wind velocity vector which has a component in the x direction, but no components on the y or z directions. This means we have a westerly wind (a wind coming from the west to the east, flowing in the positive x direction as seen in Figure 2.6). This wind vector would be written as

$$\vec{u}(x, y, z, t) = f(x, y, z, t)\vec{i} + 0\vec{j} + 0\vec{k} \qquad (2.11)$$

Then the partial derivatives with respect to the x and y spatial dimensions are as shown graphically in Figure 2.7. As can be seen, if $\partial \vec{u}/\partial x$ is positive, and all other partial derivatives are zero, then the wind speed will increase as one moves to the east, but not change as one moves to the north (or up). Conversely, if $\partial \vec{u}/\partial y$ is positive and all other partial derivatives are zero, then the westerly wind speed will increase as one moves north.

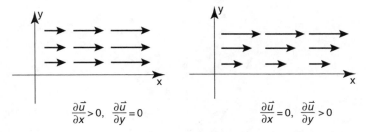

Figure 2.7 Partial derivatives of $\vec{u}$ with respect to x and y, shown only in two dimensions for clarity

2.6 Line and surface integrals

The integral of a function of a single variable is defined over some interval of the x axis such as in

$$\int_a^b f(x)\,dx$$

A natural extension to this definition is to replace this segment of the straight x axis by a general curve in space. Such an integral is called a line integral. Similarly, a double integral of a function of two variables is defined over some region of the xy plane, and it is possible to replace this region with a surface in three-dimensional space. Such an integral is called an area or surface integral. Here, we briefly introduce these two concepts.

2.6.1 Line integrals

Consider a vector field, composed of well-defined scalar functions, that represents a field of forces $\vec{F}(x, y)$ acting on the air due to the presence of a gradient in atmospheric pressure (see Chapter 4). For simplicity we initially restrict ourselves to two dimensions.

Suppose now that we have a parcel of air moving along some smooth directed curve given by the function

$$\vec{S}(t) = f(t)\vec{i} + g(t)\vec{j} \qquad (2.12)$$

which is lying in the force field (see Figure 2.8). As the parcel of air moves a short distance along the curve, we can consider that segment of the curve to be a straight line. Hence, the amount of work done on the air parcel is approximately $\vec{F} \bullet \delta \vec{S}$. That is, only the component of $\vec{F}(x, y)$ directed along the curve is doing work on the air parcel, and hence we use the scalar product. This allows us then to define the total

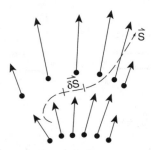

Figure 2.8 Path of an air parcel through a pressure gradient force field

amount of work done on the air parcel by the pressure gradient force, and in so doing we define the line integral:

$$\text{work} = \int_{\vec{S}} \vec{F} \bullet d\vec{S} \qquad (2.13)$$

As can be seen from this expression, the value of a line integral depends not only on the function being integrated, but also on the path taken through the space. In fact, reversing the direction of travel along the curve reverses the sign of the integral. In the case where the curve is a closed curve (that is, the starting and end points coincide), the notation is

$$\oint_{\vec{S}} \vec{F} \bullet d\vec{S}$$

For such curves, the convention is to consider the counterclockwise direction, or the direction in which the bounded region is to the left, to be the positive direction.

To extend this definition of the line integral to a more general statement in three-dimensional space, we can write

$$\int_{\vec{S}} \vec{F} \bullet d\vec{S} = \int_{\vec{S}} \left(f(x,y,z,t)\vec{i} + g(x,y,z,t)\vec{j} + h(x,y,z,t)\vec{k} \right) \bullet (dx\vec{i} + dy\vec{j} + dz\vec{k})$$

$$\int_{\vec{S}} \vec{F} \bullet d\vec{S} = \int_{\vec{S}} f dx + g dy + h dz \qquad (2.14)$$

The expression $f(x,y,z,t)dx + g(x,y,z,t)dy + h(x,y,z,t)dz$ is called an *exact differential* if there is a differentiable scalar function $\phi(x,y,z,t)$ such that

$$\frac{\partial \phi}{\partial x} = f(x,y,z,t) \quad \frac{\partial \phi}{\partial y} = g(x,y,z,t) \quad \frac{\partial \phi}{\partial z} = h(x,y,z,t)$$

When such a function exists it is called a *potential function*, or a potential of $\vec{F}$. It turns out that when such an exact differential is being integrated, the line integral in Equation (2.14) is independent of the path taken through the space, and in fact

$$\int_S \vec{F} \bullet d\vec{S} = \phi(B) - \phi(A) \tag{2.15}$$

where A is the starting point of the curve and B is the end point of the curve. This has the corollary that the line integral around a closed curve of an exact differential is identically zero.

Example For the pressure gradient force field

$$\vec{F}(x, y, z, t) = (2C_1 xy \sin C_2 t)\vec{i} + \left(C_1 x^2 \sin C_2 t + \frac{2yz}{C_3}\right)\vec{j} + \left(\frac{y^2}{C_3}\right)\vec{k}$$

where $C_1 = 0.2\,\text{N}\,\text{m}^{-3}$, $C_2 = \pi/86\,400$, $C_3 = 2\,\text{m}^3\,\text{N}^{-1}$, the work done on an air parcel traveling from location A(10,8,6) at time $t = 3\,\text{h}$ to location B(14,10,3) at time $t = 6\,\text{h}$ is

$$\text{work} = \int_S \vec{F} \bullet d\vec{S} = \int_S f dx + g dy + h dz$$

$$= \int_{10}^{14} (2C_1 xy \sin C_2 t) dx + \int_{8}^{10} \left(C_1 x^2 \sin C_2 t + \frac{2yz}{C_3}\right) dy + \int_{6}^{3} \left(\frac{y^2}{C_3}\right) dz$$

However, it can be seen that this forms an exact differential, since we can define a scalar differentiable function ϕ such that

$$\phi(x, y, z, t) = C_1 x^2 y \sin C_2 t + \frac{y^2 z}{C_3}$$

and so we can write

$$\text{work} = \phi(B) - \phi(A)$$

$$= \left[C_1 x^2 y \sin C_2 t + \frac{y^2 z}{C_3}\right]_{(14,10,3,21\,600)} - \left[C_1 x^2 y \sin C_2 t + \frac{y^2 z}{C_3}\right]_{(10,8,6,10\,800)}$$

$$= \left[0.2 \times 196 \times 10 \times 0.71 + \frac{300}{2}\right] - \left[0.2 \times 100 \times 8 \times 0.38 + \frac{384}{2}\right]$$

$$= 428.3 - 252.8$$

$$= 175\,\text{N}\,\text{m}$$

Noting that C_1 is given to one significant figure this gives an answer of $200\,\text{N}\,\text{m}$.

2.6.2 Converting line integrals to area integrals

Returning to the case of a function in two dimensions, we may choose to convert a line integral taken over the closed curve boundary of an area to a double integral of a suitable function over that area. Without showing a proof of such a relationship, we assert that it can be shown that

$$\oint_{\vec{S}} \vec{F} \bullet d\vec{S} = \oint_{\vec{S}} f dx + g dy \equiv \iint_A \left(\frac{\partial g}{\partial x} - \frac{\partial f}{\partial y} \right) dA \qquad (2.16)$$

This is known as Green's theorem, the proof of which can be found in most beginning calculus texts.

2.7 Eulerian and Lagrangian frames of reference

There are two ways we can describe the motion (or flow) of a fluid such as the Earth's atmosphere: the Eulerian and the Lagrangian. In an Eulerian frame of reference, the flow quantities such as temperature or velocity are defined as functions of position in space and time (for example, Figure 2.9). The primary flow quantity is the velocity vector field, but the complete description includes the spatial distribution of other quantities of interest such as temperature, pressure, and density. A flow variable is written as a function of position and time, $F(x, y, z, t)$, and the partial derivative gives only the local rate of change at a particular location and time. So, for example, in Figure 2.9, the temperature of the smoke flowing from a chimney can be expressed using an Eulerian frame of reference as a function $T(x, y, z, t)$, and the temperature at location O will be $T_O = T(x_O, y_O, z_O, t)$.

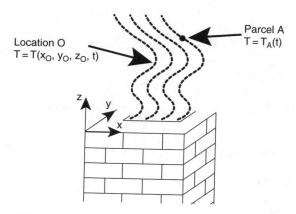

Figure 2.9 Eulerian (O) and Lagrangian (A) descriptions of the temperature of smoke from a chimney

Typically, in the Eulerian description, the components of the wind field are designated $\vec{u} = u(x, y, z, t)\vec{i} + v(x, y, z, t)\vec{j} + w(x, y, z, t)\vec{k}$ or simply $\vec{u} = (u, v, w)$. This standard usage for wind components will be used in the rest of this book.

The Lagrangian specification makes use of the fact that, as in particle mechanics, some of the dynamical and physical quantities refer not only to certain positions in space but also (and more fundamentally) to identifiable pieces of matter. The flow quantities here are defined as functions of time and the choice of the piece of matter, or *parcel*, and thus describe the dynamical history of the selected parcel. In this description then, any flow variable (including the location of a parcel) is expressed as a function of time only, $F(t)$. Since parcels change shape as they move, parcels must be chosen such that they are considered to be 'small', and that 'smallness' must continue throughout time. So, for example, in Figure 2.9, the temperature of a parcel of smoke flowing from the chimney can be expressed using a Lagrangian specification as a function of time $T(t)$, and the temperature of parcel A will be $T_A(t)$.

The Lagrangian description is useful in some contexts, such as the tracking of air pollution, and may appear to be simpler. However, it can become cumbersome when there are many parcels to be tracked, such as within a large cyclone.

2.8 Advection

The way we describe the changes and motions occurring in the atmosphere is through conservation laws:

- conservation of mass;
- conservation of momentum;
- conservation of energy.

These conservation laws apply to parcels of air in the same way that they apply to individual bodies or particles. However, in order to apply these conservation laws in an Eulerian frame, we must determine how forces (like gravity) and processes (like heating from condensation or radiation) change the values of quantities such as wind speed and direction or temperature at fixed points in space. In the Eulerian frame, quantities do not change only as a result of forces and processes acting on the parcel; change at a given point can occur simply because one air parcel has moved on from that point in the frame and another air parcel has replaced it.

Consider an air parcel with a temperature T experiencing a steady southerly wind as shown in Figure 2.10 in the left hand panel. Let us assume for now that there is no heating or cooling and hence in Lagrangian terms

$$\frac{DT}{Dt} = 0 \qquad (2.17)$$

since there are no processes acting on the air parcels that can change the temperature.

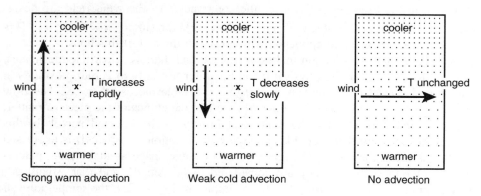

Figure 2.10 Temperature change at a point in the presence of a strong southerly, a weak northerly, and a strong westerly wind

However, in Eulerian terms there will be a change in temperature irrespective of heating or cooling, since at this location, warmer air is replacing the cooler air that is flowing northward. This is called *warm advection*, and is what you experience when, in the Northern Hemisphere, you feel a warm southerly wind, for example.

If the wind were flowing in the other direction, as shown in Figure 2.10 in the center panel, then the temperature would decrease as cooler air parcels replaced warmer ones. This is known as *cold advection*.

The rate at which this advection occurs must depend on two things – the speed of the wind that is transporting the air parcels with differing temperatures, and the strength of the temperature gradient. Also important is the orientation of the temperature gradient with respect to the wind direction – if the wind is flowing along isotherms (lines of constant temperature) then no advection will take place (Figure 2.10, right hand panel).

Any quantity can be advected, although it is most easily understood in the case of temperature. Mass (in the form of density), momentum (in the form of the wind velocity vector), and water vapor can all be modified at a particular location by this process.

Now we turn to the mathematical formulation of advection. Deriving the advective rate of change of any quantity requires a relationship between:

- the rate of change of the quantity at a fixed point, the local or Eulerian derivative $\partial/\partial t$, and

- the rate of change of the quantity following the motion, the substantial, material, or Lagrangian derivative D/Dt.

Consider again the temperature. For a given air parcel, the Lagrangian rate of change of temperature DT/Dt is only a function of time. However, at a particular point in the

Eulerian frame, the change in temperature is no longer just a function of time, but also of position in the fluid:

$$\frac{DT}{Dt} = \lim_{\delta t \to 0} \frac{\delta T}{\delta t}$$

$$\delta T = \left(\frac{\partial T}{\partial t}\right)\delta t + \left(\frac{\partial T}{\partial x}\right)\delta x + \left(\frac{\partial T}{\partial y}\right)\delta y + \left(\frac{\partial T}{\partial z}\right)\delta z$$

$$\Rightarrow \frac{DT}{Dt} = \frac{\partial T}{\partial t} + \frac{\partial T}{\partial x}\frac{Dx}{Dt} + \frac{\partial T}{\partial y}\frac{Dy}{Dt} + \frac{\partial T}{\partial z}\frac{Dz}{Dt}$$

$$= \frac{\partial T}{\partial t} + u\frac{\partial T}{\partial x} + v\frac{\partial T}{\partial y} + w\frac{\partial T}{\partial z}$$

$$\frac{\partial T}{\partial t} = \frac{DT}{Dt} - \left(u\frac{\partial T}{\partial x} + v\frac{\partial T}{\partial y} + w\frac{\partial T}{\partial z}\right) \quad (2.18)$$

This relationship gives the rate of advective temperature change $-(u\partial T/\partial x + v\partial T/\partial y + w\partial T/\partial z)$. The sign for the temperature advection makes physical sense, since the wind must be blowing down the temperature gradient for there to be a temperature increase. For example, in Figure 2.10, left panel, the wind vector is positive in the y direction, and the temperature gradient is negative (that is, it gets colder) in the y direction. This configuration yields an advective temperature *increase*.

Example The air at a point 50 km north of a station is 3 °C cooler than at the station (Figure 2.11). If the wind is blowing from the north-east at $20\,\mathrm{m\,s^{-1}}$ and the air at the station is being heated by radiation at the rate of $1\,°\mathrm{C\,h^{-1}}$, what is the temperature change at the station?

The temperature change at the station is

$$\left.\frac{\partial T}{\partial t}\right|_{station} = \left.\frac{\partial T}{\partial t}\right|_{heating} + \left.\frac{\partial T}{\partial t}\right|_{advection}$$

$$= \frac{DT}{Dt} - \left(u\frac{\partial T}{\partial x} + v\frac{\partial T}{\partial y} + w\frac{\partial T}{\partial z}\right)$$

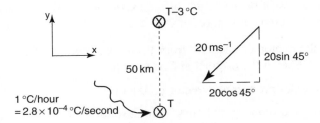

Figure 2.11 Station with temperature T undergoing radiative heating and advective cooling

$$\frac{DT}{Dt} = \frac{1\,°C}{3600\,s} = 3 \times 10^{-4}\,°C\,s^{-1}$$

$$\vec{u} = -14\vec{i} - 14\vec{j}$$

$$\frac{\partial T}{\partial x} = 0 \qquad \frac{\partial T}{\partial y} = \frac{-3\,°C}{50\,km} = -6 \times 10^{-5}\,°C\,m^{-1} \qquad \frac{\partial T}{\partial z} = 0$$

$$\left.\frac{\partial T}{\partial t}\right|_{station} = 3 \times 10^{-4} - (-14 \times -6 \times 10^{-5})$$

$$= -5 \times 10^{-4}\,°C\,s^{-1}$$

Hence, in this case, the cooling by advection overwhelms the warming from the Sun, and the temperature at the station gets lower.

Review questions

2.1 To an observer on the ground, the wind speed experienced as a hurricane passes overhead is equal to the wind speed due to the circulation of air around the hurricane plus the speed of movement of the hurricane. Consider a Northern Hemisphere hurricane with a wind speed of $50\,m\,s^{-1}$ blowing in a counterclockwise direction around the center of the hurricane at a distance of 10 km from the center. Assume that the center of the hurricane is moving towards the north at $5\,m\,s^{-1}$.

(a) Sketch the wind vectors associated with the circulation of the hurricane only, at two points 10 km due east and due west from the center of the hurricane.

(b) Sketch the vector that represents the motion of the center of the hurricane.

(c) What is the wind speed experienced by an observer who is 10 km due east of the center of the hurricane? What is the wind speed 10 km due west of the center of the hurricane? Determine your answers graphically.

(d) Write the three vectors you sketched in parts (a) and (b) in component form, using the coordinate system shown in Figure 2.6.

(e) Use Equations (2.1) and (2.2) to calculate the wind speed experienced by an observer who is 10 km due east of the center of the hurricane and one who is 10 km due west of the center of the hurricane. Do these results match your graphic solutions in part (c)?

2.2 Calculate the vector that results from the scalar product of $\vec{u} = 10\vec{i} + 3\vec{j} - 6\vec{k}$ and $c = 1.5$. Sketch the original and new vectors.

2.3 Calculate the dot product and vector product of:

(a) an east wind blowing at $10\,m\,s^{-1}$ and a south wind blowing at $2\,m\,s^{-1}$;

(b) a south wind blowing at $4\,m\,s^{-1}$ and a north wind blowing at $6\,m\,s^{-1}$.

2.4 Numerical weather prediction models that are used to forecast the weather often use coordinate systems that are rotated relative to the one shown in Figure 2.6. Given the wind vector $\vec{u} = u_x \vec{i} + u_y \vec{j}$, where $u_x = 10\,\mathrm{m\,s^{-1}}$ and $u_y = 0\,\mathrm{m\,s^{-1}}$, determine the wind vector in a coordinate system rotated:

(a) 30° counterclockwise

(b) 30° clockwise

(c) 90° counterclockwise

(d) 180°

relative to the coordinate system used by meteorologists.

2.5 Calculate the partial derivatives in the x, y, and z directions of the vector

$$\vec{u}(x, y, z, t) = (t^3 xz)\vec{i} + (4\sin xt + 5y^3)\vec{j} + (2xy^2)\vec{k}$$

which was used in the example in Section 2.5.

2.6 Using the horizontal wind vector

$$\vec{u}(x, y, z, t) = \left[C \sin\left(\frac{2\pi}{L} x\right) + \frac{C}{4000} y \right] \vec{i}$$

where $C = 10\,\mathrm{m\,s^{-1}}$ and $L = 1 \times 10^6\,\mathrm{m}$:

(a) Plot the $\vec{i}$ component of $\vec{u}$ for $-1000\,\mathrm{km} \le x \le 1000\,\mathrm{km}$ at $y = -500$, 0, and 500 km.

(b) Calculate $\partial \vec{u}/\partial x$.

(c) Plot $\partial \vec{u}/\partial x$ for $-1000\,\mathrm{km} \le x \le 1000\,\mathrm{km}$.

(d) Describe the relationship between $\vec{u}$ and $\partial \vec{u}/\partial x$ shown in the plots from parts (a) and (c). Is this consistent with your understanding of the physical meaning of $\partial \vec{u}/\partial x$?

2.7 Using the surface weather map in Figure 1.7 calculate the rate of change of temperature due to advection halfway between Holdrege, Nebraska and Dodge City, Kansas. Assume that the wind speed and direction at this point are equal to the average wind speed and direction reported at Holdrege and Dodge City. (Hint: the distance along the western border of Kansas is equal to 333 km.)

3 Properties of fluids

3.1 Solids, liquids, and gases

Imagine that you are holding a brick in front of you between your palms, and you try to move your right hand away from you and your left hand toward you. The forces you are exerting with your hands are transmitted through the brick from one layer of atoms to the next, in a different direction on each side of layer. In this way, you are imposing a *shear stress* on the brick. Stress is a measure of the internal forces in a body between its constituent particles, as they resist tension and compression (normal stress) or sliding (shear stress) in response to externally applied forces.

However, a brick is a solid material – it possesses the property of *rigidity*. When you apply a moderate shear stress, the brick will deform only slightly, and return to its original shape when the stress is removed. If you are a body builder and can apply a large enough stress to the brick, it will shatter rather than deform.

Now, apply this same stress to a brick made of, say, modeling clay. This is a *plastic* material, which means it will deform continuously and irreversibly. What if the brick was made of rubber? This is an *elastic* substance that will also deform continuously but will then return to its original shape. Of course, all of these materials also possess some amount of rigidity – they maintain their identity under the application of the stress.

In contrast, the defining property of a fluid is that it is a substance which deforms continuously when acted on by a shearing stress of *any* magnitude, however small. So an ideal fluid is a material with *no rigidity at all*. Liquids and gases are both considered to be fluids because they possess this defining property. The molecular mechanism by which a liquid resists deformation is not the same as that in a gas, although the governing equation determining the rate of change of deformation has the same form. A fluid which transmits shear stresses internally is called a *viscous* fluid.

Solids and fluids (both liquids and gases) also differ in the way they respond to *normal stresses*, that is stresses at right angles to the surface of the material. While a solid can support both tensile and compressive stress, a fluid usually supports only compression.

Applied Atmospheric Dynamics Amanda H. Lynch, John J. Cassano
© 2006 John Wiley & Sons, Ltd

Table 3.1 Some properties of materials

	Gas	Liquid	Solid
Preferred shape	No	Minimal	Yes
Density	Lowest	Medium	Highest
Intermolecular forces	Weak	Medium	Strong
Differential motion	Yes, with minimum force	Yes, with moderate force	Minimal, with great force
		Thixotropic substances Solids in plastic state Viscoelastic solids	Elastic solids
	No memory	Partial memory	Perfect memory
	Compressible		Incompressible

As illustrated in Table 3.1, there are many odd substances that do not neatly fit into the categories of liquids or solids, but lie on the boundary between them. For example, thixotropic substances like jelly or paint behave as an elastic solid until subjected to severe distortion such as shaking, upon which they behave more like a liquid. Viscoelastic substances like egg white have partial memory, halfway between an elastic solid and a viscous fluid. Other substances, like slurries such as toothpaste, will behave as a solid if the shear stress is small, but will behave as a fluid once a critical stress level is reached. Sea ice is considered to be 'elastic plastic' material but is often represented in computer models, moderately successfully, as a 'viscous plastic' material. The study of the properties of such unusual materials is called *rheology*.

3.2 Thermodynamic properties of air

If two samples of air can exist in contact with each other without a change in properties, the two samples are said to have the same *state*. These properties include the temperature, density, chemical composition, pressure, among others. The state of such a system can be defined completely by some subset of such properties. For an *ideal gas* this subset, known as the *state variables*, comprises any two state variables that do not depend on the mass of the system (which are called *intensive variables*).

Equations of state relate properties of state to one another. Such equations are usually empirical (that is, determined from the results of experiments). The equation of state for an ideal gas, the *Ideal Gas Law*, is given by

$$\frac{p}{\rho T} = R, \text{a constant} \tag{3.1}$$

where p is the pressure, ρ is the density, and T is the absolute temperature. The volume occupied by a mass of gas is also proportional to that mass, and hence we expect the constant R to be specific to a given chemical species. Experiment shows that at low enough densities, R is a ratio between a universal constant $R^* = 8.314\,\mathrm{J\,mol^{-1}\,K^{-1}}$ and the molecular mass M of the gas ($R = R^*/M$).

3.3 Composition of the atmosphere

The atmosphere consists of a number of different gases, each with different properties. The percentage of a volume of the atmosphere occupied by most of these gases is nearly constant as one moves horizontally through the lower portion of the atmosphere. One important exception to this is water vapor, which can vary dramatically in both the horizontal and vertical direction. Therefore it is often easiest to consider first the dry atmosphere, excluding water vapor, before looking at the moist atmosphere.

3.3.1 The dry atmosphere

The major constituents of the dry atmosphere are nitrogen and oxygen, which together make up 99% of the lower atmosphere by volume. Trace amounts of carbon dioxide, argon, ozone, and methane make up much of the remainder.

While the atmosphere is a mixture of various gases, its state can nevertheless be approximated using the ideal gas law. To do this, first note that the individual constituents of air obey this law. Hence,

$$p_n = \rho_n R_n T \tag{3.2}$$

where n represents the nth constituent.

In addition, the mixture of air obeys *Dalton's law*, which states that the total pressure exerted by a mixture of gases that do not interact chemically is simply the sum of the partial pressures of each constituent. The *partial pressure* of each constituent is the pressure it would exert if it was present alone and occupied the same volume as the whole mixture. Dalton's law can be expressed mathematically as

$$p = T \sum_n \rho_n R_n \tag{3.3}$$

If we define a weighted average molecular mass for dry air to be that based on measurements of the average composition of the dry atmosphere, $M_d = 28.966\,\mathrm{g\,mol^{-1}}$, the gas constant for dry air is given by

$$R_d = \frac{\sum_n \rho_n R^*/M_n}{\rho} = \frac{\sum_n \rho_n R_n}{\rho} = \frac{R^*}{M_d} = 287.04\,\mathrm{J\,kg^{-1}\,K^{-1}} \tag{3.4}$$

This yields an ideal gas law for the mixture of gases that make up dry air:

$$p = \rho R_d T \tag{3.5}$$

Example Calculate the pressure exerted by dry air that has a density of 1.2 kg m^{-3} at a temperature of 20°C. Remember that all units should be SI, so the temperature needs to be converted from °C to K.

$$p = \rho R_d T$$
$$= 1.2 \text{ kg m}^{-3} \times 287.04 \text{ J kg}^{-1} \text{ K}^{-1} \times (20 + 273) \text{K}$$
$$= 100\,923 \text{ J m}^{-3} = 100\,923 \text{ N m}^{-2}$$
$$= 1009 \text{ hPa}$$

3.3.2 The moist atmosphere

Unlike the other gases that make up the atmosphere, water vapor makes up a highly variable percentage by volume of the atmosphere because of the range of conditions within it. Indeed, the amount of water vapor can range from close to zero to 4% of the total volume. Additionally water vapor can undergo phase changes to both the liquid and solid state, and water can exist simultaneously as a gas, liquid, and solid in a given volume of the atmosphere. Changes of phase of water in the atmosphere release or absorb large quantities of heat which can be important in driving the dynamics of the atmosphere.

Given the unique, and important, properties of water in the atmosphere, it should come as no surprise that atmospheric scientists use a number of different variables to describe the amount of water vapor contained in the atmosphere. Some of these moisture variables include vapor pressure, specific humidity, water vapor mixing ratio, and dew point temperature.

The partial pressure of the water vapor in a given volume of the atmosphere is referred to as the *vapor pressure* (e), and is simply the pressure exerted by the water vapor molecules present in the volume of the atmosphere that is being considered.

Specific humidity and water vapor mixing ratio are closely related, and are often confused by both students and more experienced atmospheric scientists. *Specific humidity* (q) is defined as the mass of water vapor contained in a unit mass of air, while the *water vapor mixing ratio* (r) is defined as the mass of water vapor contained in a unit mass of *dry* air. Since the mass of water vapor contained in a unit mass of air is small (ranging from close to zero in the polar regions or in the upper troposphere to approximately 0.04 kg in tropical locations) the numerical values of specific humidity and mixing ratio rarely differ by more than a few percent.

Just as in the case of a dry atmosphere, we can apply the ideal gas law to water vapor. Of course water vapor has a lower molecular weight, and hence a larger gas constant, than dry air. Also, since the composition of water vapor varies considerably throughout the atmosphere, the approach of defining an average gas constant based on weighted average molecular weights is likely to yield inaccurate results for moist air. An alternative approach is to retain the gas constant for dry air as follows:

$$p = (\rho_d R_d + \rho_v R_v)T$$

$$= \rho R_d \left(\frac{\rho_d}{\rho} + \frac{\rho_v}{\rho}\frac{R_v}{R_d}\right)T$$

$$= \rho R_d \left((1-q) + q\frac{R_v}{R_d}\right)T$$

$$p = \rho R_d T_v \tag{3.6}$$

where we have defined the new variable T_v, called the *virtual temperature*, and have used the definition of specific humidity, q, to replace ρ_v/ρ. The virtual temperature is a fictitious temperature that by definition gives the temperature dry air would have if it had the same density as the moist air in question at the same pressure. It can be calculated from

$$T_v = \left((1-q) + q\frac{R_v}{R_d}\right)T \tag{3.7}$$

Since moist air is less dense than dry air, the virtual temperature is always greater than the actual temperature. However, even for very warm, moist air, the difference between the actual and the virtual temperature is just a few percentage points. Thus, for most applications, it is sufficient to use the ideal gas equation as expressed in Equation (3.5).

Example Calculate the virtual temperature of 25°C air when the water vapor mixing ratio is 19 g kg^{-1}.

For this example we will make the assumption that the specific humidity is equal to the water vapor mixing ratio; that is, we are assuming the mass of water vapor per unit mass of air is small. Then, converting all values to SI units,

$$T_v = \left((1-q) + q\frac{R_v}{R_d}\right)T$$

$$= \left((1 - 0.019\,\text{kg kg}^{-1}) + 0.019\,\text{kg kg}^{-1}\frac{461\,\text{J kg}^{-1}\,\text{K}^{-1}}{287\,\text{J kg}^{-1}\,\text{K}^{-1}}\right)298\,\text{K}$$

$$= 1.0115 \times 298$$

$$= 301\,\text{K}$$

$$T_v = 28\,°\text{C}$$

3.4 Static stability

Consider a parcel of air with temperature T and pressure p. Suppose that it is given a small amount of heat dQ per unit mass, and as a consequence its temperature and pressure change by amounts dT and dp respectively. That is, the parcel reacts to the heating though a combination of temperature rise and expansion. The heating can be imparted through conduction, convection (that is, hot air rising), or latent heat release. Then, using the ideal gas law:

$$p = \rho R_d T$$
$$p\alpha = R_d T$$
$$\Rightarrow p\frac{d\alpha}{dt} + \alpha\frac{dp}{dt} = R_d\frac{dT}{dt}$$

where $\alpha(=1/\rho)$ is called the *specific volume* of the parcel of air. Using the fact that the gas constant for dry air can be shown to be given by the difference between the specific heat capacities at constant pressure and volume respectively, the relationship $R_d = c_p - c_v$ gives

$$p\frac{d\alpha}{dt} + \alpha\frac{dp}{dt} = c_p\frac{dT}{dt} - c_v\frac{dT}{dt}$$
$$p\frac{d\alpha}{dt} + c_v\frac{dT}{dt} = c_p\frac{dT}{dt} - \alpha\frac{dp}{dt}$$

Since heating is manifest as a combination of expansion at constant temperature and increase of temperature at constant volume,

$$\frac{dQ}{dt} = c_p\frac{dT}{dt} - \alpha\frac{dp}{dt} \tag{3.8}$$

This is the *first law of thermodynamics* as applied to a parcel of air.

As we introduced in Section 2.8, the atmosphere is most usefully described using a set of conservation laws, which require the identification of conserved quantities, such as mass and momentum. In this case, we can use the first law of thermodynamics to derive a new thermodynamic variable which is conserved under certain conditions. This variable is known as the *potential temperature*, which is the temperature a parcel of air would possess if it were brought to a given reference pressure with no heat exchange with the surroundings. Such a process is called an *adiabatic* process and the reference pressure is typically taken to be 1000 hPa. For adiabatic motions, that is, motions wherein no heat is exchanged with the surrounding environment, potential temperature is a conserved quantity. Temperature may not be conserved in such motions.

First, we use the Ideal Gas Law to eliminate the specific volume from Equation (3.8),

$$dQ = c_p dT - \frac{R_d T}{p} dp$$

STATIC STABILITY

In an adiabatic process, $dQ = 0$, and hence we can integrate the above expression,

$$0 = \frac{c_p}{R_d T} dT - \frac{1}{p} dp$$

$$C = \frac{c_p}{R_d} \ln T - \ln p$$

where C is a constant of integration. Applying the boundary condition that T is equal to the potential temperature, denoted θ, when $p_0 = 1000\,\text{hPa}$,

$$C = \frac{c_p}{R_d} \ln \theta - \ln p_0$$

$$\frac{c_p}{R_d} \ln \theta - \ln p_0 = \frac{c_p}{R_d} \ln T - \ln p$$

$$\ln \left(\frac{\theta^{c_p/R_d}}{T^{c_p/R_d}} \right) = \ln \left(\frac{p_0}{p} \right)$$

$$\Rightarrow \theta = T \left(\frac{p_0}{p} \right)^{R_d/c_p} \tag{3.9}$$

Example Calculate the potential temperature of an air parcel that has a temperature of 10°C and a pressure of 850 hPa.

For this example we will use $p_0 = 1000\,\text{hPa}$, $R_d = 287\,\text{J kg}^{-1}\,\text{K}^{-1}$, and $c_p = 1004\,\text{J kg}^{-1}\,\text{K}^{-1}$. The temperature must be converted to absolute units K.

$$\theta = 283 \left(\frac{1000 \times 10^2}{850 \times 10^2} \right)^{287/1004}$$

$$\theta = 297\,\text{K}$$

$$\theta = 23\,°\text{C}$$

In atmospheric science conserved quantities such as potential temperature are useful since they allow us to track the movement of air, while other non-conserved properties of the air are changing. From Equation (3.9) we note that an air parcel that is displaced vertically in an adiabatic process will experience a temperature decrease as the air parcel is lifted (p decreases while θ remains constant), and this cooling is referred to as *adiabatic cooling*. Conversely, an air parcel that is forced to sink adiabatically (p increases) will warm in a process referred to as *adiabatic warming*.

Example What will the temperature of the air parcel from the example above be if the air parcel is now lifted from 850 hPa to 500 hPa, assuming that no phase change of water occurs as the parcel is lifted?

From the previous example we know that the potential temperature of this air parcel is 297 K or 23 °C. Rearranging Equation (3.9) gives the following:

$$T = \theta \left(\frac{p}{p_0}\right)^{R_d/c_p}$$

$$= 297 \left(\frac{500 \times 10^2}{1000 \times 10^2}\right)^{287/1004}$$

$$= 244 \, \text{K}$$

$$T = -29 \, °\text{C}$$

The vertical derivative of the atmospheric potential temperature profile is also a fundamental quantity for characterizing the stability of air parcels undergoing vertical motion. Consider an air parcel with volume V that starts at height z and is lifted adiabatically to height specified to be $z + dz$. The initial potential temperature of the air parcel and the environment are equal at height z. Therefore the temperature of both the air parcel and the environment at height z will also be equal and will be given by

$$T_z = \theta_z \left(\frac{p_z}{p_0}\right)^{R_d/c_p}$$

As the air parcel is lifted to height $z + dz$ its temperature will now be given by

$$T_{parcel,z+dz} = \theta_z \left(\frac{p_{z+dz}}{p_0}\right)^{R_d/c_p}$$

Note that the potential temperature of the air parcel has not changed between heights z and $z + dz$ since the air parcel is being lifted adiabatically and potential temperature is a conserved quantity for adiabatic processes.

The temperature of the air parcel's environment at height $z + dz$ will be given by

$$T_{z+dz} = \theta_{z+dz} \left(\frac{p_{z+dz}}{p_0}\right)^{R_d/c_p}$$

where the potential temperatures of the environment at heights z and $z + dz$ are not assumed to be equal and are given by θ_z and θ_{z+dz} respectively.

The *buoyancy force* or upward thrust per unit mass F experienced by the parcel at $z + dz$ due to the air is

$$F = \frac{\text{(weight of air displaced)} - \text{(weight of air parcel)}}{\text{(mass of parcel)}}$$

$$= \frac{g\rho_{z+dz}V - g\rho_{parcel,z+dz}V}{\rho_{parcel,z+dz}V}$$

Canceling V and using the ideal gas law, this expression becomes

$$F = g\frac{(p/R_d T_{z+dz}) - (p/R_d T_{parcel,z+dz})}{(p/R_d T_{parcel,z+dz})}$$

$$= g\frac{1/T_{z+dz} - 1/T_{parcel,z+dz}}{1/T_{parcel,z+dz}}$$

$$= g\frac{T_{parcel,z+dz} - T_{z+dz}}{T_{z+dz}}$$

$$= g\frac{\theta_{parcel,z+dz} - \theta_{z+dz}}{\theta_{z+dz}}$$

Using the fact that the potential temperatures of the air parcel at heights z and $z+dz$ are equal we can replace $\theta_{parcel,z+dz}$ with θ_z to give

$$F = g\left(\frac{\theta_z - \theta_{z+dz}}{\theta_{z+dz}}\right)$$

$$= -\frac{g}{\theta}\frac{d\theta}{dz}dz$$

$$= -N^2 dz$$

where we have defined the *buoyancy frequency*, N, which is also known as the *Brunt–Väisälä frequency*. This is the frequency at which a parcel will oscillate if displaced vertically and acted upon by the restoring force arising from the buoyancy of the parcel.

If the potential temperature of the environment is uniform with height, the displaced parcel experiences no buoyancy force and will remain at its location. Such a layer of air is said to be *neutrally stable*. If the potential temperature of the environment increases with height, a parcel displaced upward experiences a negative restoring force, and vice versa, and hence will tend to return to its equilibrium level. Thus $d\theta/dz > 0$ of the environment characterizes a stable layer of air. In contrast, if the potential temperature of the environment were to decrease with height, a displaced parcel would experience a force in the direction of the displacement and continue to accelerate in the direction of the displacement, clearly an unstable situation.

The above discussion assumes that no phase change of water occurs in the air parcel as it is displaced, and this type of process is referred to as a *dry adiabatic* process. It is important to note that a dry adiabatic process does not refer to a process in which water is not present, only that the water that is present (in any of the vapor, liquid, or solid phases) does not change phase during the process.

If the water in an air parcel (vapor, liquid, or solid) experiences a phase change as the parcel is displaced, latent heat will be released or absorbed by the water during the phase change resulting in a change in the potential temperature of the air parcel. For this situation the air parcel is said to experience a *moist adiabatic* process. The special situation of moist adiabatic processes will be discussed further in Chapter 11 in the discussion of convective systems.

3.5 The continuum hypothesis

The science of fluid dynamics is concerned with behavior on a macroscopic scale; that is, a scale that is large compared to the distance between molecules. Because of this, it can be assumed that a fluid is perfectly continuous in structure, and physical quantities such as mass and temperature are spread uniformly over the volume of fluid. Hence, the fluctuations arising from the different properties of molecules can be assumed to have no effect on the observations at meteorological space and time scales. This is known as the *continuum hypothesis*.

What does it mean for the fluid properties to vary smoothly?

Consider a laboratory experiment in which an instrument is inserted in a fluid. The *sensitive volume* of the instrument, that is the volume over which the instrument can detect variations, must be small enough to measure the 'local' property (that is, no significant variation in that property within the volume). At the same time, the sensitive volume must be large compared to the molecular scale. In this example, let us assume that our instrument is very sensitive, with a sensitive volume of $10^{-5}\,\text{m}^3$. At normal temperatures and pressures, this volume of air would contain about 3×10^{10} molecules. This is large enough for an average over the molecules to be independent of their number. Further, the mean free path of the molecules is about $5 \times 10^{-8}\,\text{m}$, which is small enough to be contained within the volume. Hence, we know that our instrument will measure the macroscopic behavior of our fluid.

In fact, problems generally occur only when the number of molecules in a given volume is very low (such as the atmosphere at the altitude of a satellite) or if the variation across the volume is large (such as in a shock wave). Thus, the hypothesis implies that we can attach a definite and real meaning to the idea of fluid properties 'at a point', and these properties are continuous functions of position in the fluid and of time.

3.6 Practical assumptions

Most practical applications of fluid dynamics concern water or air, or fluids which closely resemble one of these (leaving aside such fluids as molten polymers, liquid crystals, or plasmas in a magnetic field). In this book, we are considering air, which is a fluid composed of a mixture of gases, including nitrogen, oxygen, and water vapor, among many others. For such a fluid we can make certain practical assumptions:

1. The atmosphere is an isotropic fluid. This means that the properties of the fluid and their spatial derivatives do not depend on direction.
2. The atmosphere is a Newtonian fluid. Strictly speaking, this means that the fluid obeys a linear relationship between shear stress and rate of deformation. We

will not address the details of this relationship in this book. Suffice to say that this gives us a simple way of treating viscous forces (that is, friction) in the atmosphere.

3. The atmosphere is a classical fluid. This means that motions in the atmosphere are governed by classical mechanics and thermodynamics, not quantum physics!

3.7 Continuity equation

In most situations important to atmospheric dynamics, mass is not created or destroyed – it is always conserved. This idea is often termed *continuity*. However, even if the mass remains constant, the volume may change – air can expand or flow outward (a process called divergence, as introduced in Section 1.4.1), or compress (*convergence*). Hence, the approach to describing this mass conservation requirement is to consider how the density changes in the presence of divergence or convergence to keep the total mass constant.

Consider a small volume of air V at some fixed point in our Eulerian frame of reference. The mass of air in that volume at any instant is simply the density multiplied by the volume. However, we cannot assume that the density is constant throughout the volume, so we express this mass as an integral of density ρ over the volume:

$$M_{air} = \iiint \rho \, dx \, dy \, dz \tag{3.10}$$

For the purposes of our argument, we will assume that our volume is a box as shown in Figure 3.1. We must assume that air is flowing through this box, which is fixed in space, all the time, so that the mass of air in the box accumulates at a rate equal to the total inflow minus the total outflow at each face of the box.

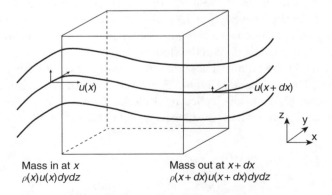

Mass in at x
$\rho(x)u(x)dydz$

Mass out at $x+dx$
$\rho(x+dx)u(x+dx)dydz$

Figure 3.1 Air flowing into and out of a box fixed in space

Hence, we can express the rate of change of mass in the box as follows:

$$\frac{\partial M_{air}}{\partial t} = \frac{\partial}{\partial t}\iiint \rho\,dx\,dy\,dz = \iint (\rho u)_x - (\rho u)_{x+dx}\,dy\,dz$$
$$+ \iint (\rho v)_y - (\rho v)_{y+dy}\,dx\,dz$$
$$+ \iint (\rho w)_z - (\rho w)_{z+dz}\,dx\,dy$$

This can then be simplified to give the *continuity equation*:

$$\frac{\partial}{\partial t}\iiint \rho\,dx\,dy\,dz = -\iiint \frac{\partial(\rho u)}{\partial x} + \frac{\partial(\rho v)}{\partial y} + \frac{\partial(\rho w)}{\partial z}\,dx\,dy\,dz$$

$$\frac{\partial \rho}{\partial t} + \frac{\partial(\rho u)}{\partial x} + \frac{\partial(\rho v)}{\partial y} + \frac{\partial(\rho w)}{\partial z} = 0 \qquad (3.11)$$

The terms ρu, ρv, and ρw are called mass *fluxes*, that is density multiplied by speed of flow. Hence, this form of the continuity equation is called the 'flux form' because it is making use of the gradients of the mass fluxes. We can expand the derivatives of the mass fluxes to arrive at a different form of the continuity equation:

$$\underbrace{\frac{\partial \rho}{\partial t}}_{(1)} + \underbrace{\left(u\frac{\partial \rho}{\partial x} + v\frac{\partial \rho}{\partial y} + w\frac{\partial \rho}{\partial z}\right)}_{(2)} + \rho\underbrace{\left(\frac{\partial u}{\partial x} + \frac{\partial v}{\partial y} + \frac{\partial w}{\partial z}\right)}_{(3)} = 0 \qquad (3.12)$$

Term (2) represents the net transport of mass into and out of the volume. This term accounts for the variation in density across the volume while holding the flow constant. Hence, terms (1) and (2) together represent the Lagrangian rate of change in density, which is made up of the local (Eulerian) rate of change at a point, minus the advection of density (see Section 2.7). Term (3) represents the *divergence* of the flow while holding density constant. For example, considering only the x component of the divergence, if $\partial u/\partial x > 0$, then the flow is getting faster as we move to the east. Hence, air is leaving the volume faster than it is entering, contributing to a decrease in mass in the volume (see also Figure 2.7). The form of the continuity equation shown in Equation (3.12) is also called the Eulerian form of the equation.

Consider an air column where the density is not changing at all with time, so that

$$\frac{\partial \rho}{\partial t} = 0$$

$$\Rightarrow \frac{\partial(\rho u)}{\partial x} + \frac{\partial(\rho v)}{\partial y} + \frac{\partial(\rho w)}{\partial z} = 0$$

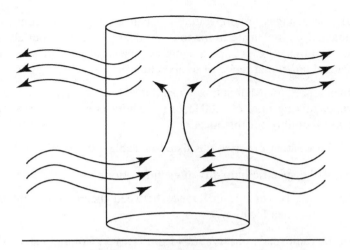

Figure 3.2 An illustration of Dines' compensation, showing convergence at a low level and divergence at an upper level of a column. This flow is typical of the low-level circulation in cyclone

If we also assume that the density is approximately constant in space (clearly a very strong assumption), then we have

$$\rho\left(\frac{\partial u}{\partial x} + \frac{\partial v}{\partial y} + \frac{\partial w}{\partial z}\right) = 0$$

$$\Rightarrow \frac{\partial u}{\partial x} + \frac{\partial v}{\partial y} = -\frac{\partial w}{\partial z}$$

This relationship, known as Dines' compensation, suggests that in order that there be no net change in mass in the column, any horizontal or vertical divergence of air in the column is replaced by the convergence of air at other levels in the column (Figure 3.2). In the real atmosphere, convergence does not always exactly balance divergence, but, in fact, departures from this balance are generally quite small.

Review questions

3.1 Using the ideal gas law for dry air calculate the density of air when

(a) the temperature is 0°C and the pressure is (i) 1000 hPa, (ii) 500 hPa, and (iii) 300 hPa; and

(b) the pressure is 1000 hPa and the temperature is (i) −30°C, (ii) 0°C, and (iii) 30°C.

(c) Based on your answers to parts (a) and (b), are the changes in density greater for changes in pressure at constant temperature or for changes in temperature at a constant pressure?

3.2 Typically, do changes in temperature or changes in pressure contribute more to observed changes in density in the mid-latitudes? To guide your answer, assume that a sea level location is experiencing an observed range of pressure of 980 to 1040 hPa and an annual range of temperature of $-30\,°C$ to $+30\,°C$.

3.3 A weather report for Melbourne, Australia indicates a temperature of $15\,°C$, a water vapor mixing ratio of $0.003\,\text{kg}\,\text{kg}^{-1}$, and a pressure of 1000 hPa. What is the specific humidity for this location?

3.4 Consider the weather observations listed in Table 3.2.

 (a) What is the specific humidity at each location?

 (b) What is the percentage difference between the mixing ratio and specific humidity at each location?

3.5 Consider the radiosonde observations from Denver, Colorado at 12 UTC 29 Sep 2005, listed in Table 3.3.

 (a) What is the specific humidity at each pressure level?

 (b) What is the percentage difference between the mixing ratio and the specific humidity at each of these pressure levels?

3.6 Calculate the virtual temperature for each location listed in Table 3.2. What is the percentage difference between the temperature and the virtual temperature at each of these locations?

3.7 Calculate the potential temperature at each pressure level listed in Table 3.3.

Table 3.2 Meteorological observations for calculations required in question 3.4

Location	Temperature	Mixing ratio	Pressure
Belem, Brazil	$31\,°C$	$22\,\text{g}\,\text{kg}^{-1}$	1009 hPa
New York, NY	$23\,°C$	$10\,\text{g}\,\text{kg}^{-1}$	1005 hPa
Barrow, AK	$-3\,°C$	$2.5\,\text{g}\,\text{kg}^{-1}$	1011 hPa

Table 3.3 Meteorological observations for calculations required in question 3.5

Pressure level	Height	Temperature	Mixing ratio
843 hPa	1625 m	$5\,°C$	$6.2\,\text{g}\,\text{kg}^{-1}$
500 hPa	5840 m	$-9\,°C$	$0.8\,\text{g}\,\text{kg}^{-1}$
300 hPa	9570 m	$-38\,°C$	$0.3\,\text{g}\,\text{kg}^{-1}$
100 hPa	16520 m	$-64\,°C$	$0.01\,\text{g}\,\text{kg}^{-1}$

REVIEW QUESTIONS

3.8 Calculate the Brunt–Väisälä frequency using the data in Table 3.3 for the layers:

(a) 843 to 500 hPa

(b) 500 to 300 hPa

(c) 300 hPa to 100 hPa.

3.9 An air parcel that has a temperature of 17°C at the 850 hPa standard level is lifted dry adiabatically. What is the density of the parcel when it reaches 500 hPa?

3.10 Consider an air parcel with an initial pressure of 1000 hPa and a temperature of 20°C. This air parcel rises dry adiabatically over a mountain range to a pressure of 700 hPa and then descends on the downwind side of the mountain range to a pressure of 850 hPa.

(a) What is the temperature, potential temperature, and density of this air parcel at the initial position, at the top of the mountain range, and on the downwind side of the mountain range?

(b) How will the temperature, potential temperature, and density of this air differ from that found in part (a) if condensation of water vapor occurs as the air parcel rises to the top of the mountain range. (Hint: a qualitative answer for this question is acceptable.)

3.11 (a) How will the potential temperature of an air parcel change if, as it ascends adiabatically, the water vapor condenses to form a cloud?

(b) In this situation, is the criterion $d\theta/dz > 0$ sufficient to identify a stable layer of the atmosphere? Why or why not?

3.12 Surface weather observations across an east–west-oriented cold front indicate winds from the north at $10\,\mathrm{m\,s^{-1}}$ on the north side of the cold front and winds from the south at $5\,\mathrm{m\,s^{-1}}$ on the south side of the front. These weather observations are made at locations that are 100 km apart. Assume that these winds occur over the lowest 1 km of the atmosphere and that the vertical velocity at the surface is $0\,\mathrm{m\,s^{-1}}$. Calculate the vertical velocity at a height of 1 km assuming that the density of the air does not vary in time or space.

4 Fundamental forces

4.1 Newton's second law: $F = ma$

Because we have made the assumptions in Chapter 3 that the atmosphere is a classical Newtonian fluid, continuous in structure and physical properties at the scales of interest, we can simply use Newton's second law of motion to describe the acceleration $\vec{a}$ of the atmosphere. However, our approach will differ from that of the classical mechanics of solid bodies in that we must achieve this description for a continuous field of matter with varying density, rather than for discrete objects with constant mass. Hence, it is most straightforward to write this description as

$$\frac{\text{force}}{\text{mass}} = \vec{a} \qquad (4.1)$$

By using this relationship and predicting the distribution of density based on the continuity equation (Equation (3.3)), we can in theory determine the field of acceleration for any situation. So, in order to examine the motion of the atmosphere, we need to consider all forces that are acting upon it.

4.2 Body, surface, and line forces

The forces that act upon a fluid in general can be split into two types:

- body or volume forces, which act at a distance and affect the entire fluid volume; and
- surface forces, which act locally upon a part of a fluid.

Body forces include gravity, electromagnetic forces, and forces that arise due to the acceleration of the frame of reference. In all of these cases, the medium is in contact with a force field of some sort.

A surface force is one in which there is direct mechanical contact between two bodies. In a fluid, that is a gas or a liquid, the situation is more complex than in the case of a solid. In a gas this contact is due predominantly to transport of momentum across the boundary by migrating molecules. In a liquid there are additional contributions arising from forces between molecules on either side of the boundary. The primary surface force of interest in the study of atmospheric dynamics is stress. The boundary across which the stress is acting need not be the physical interface between two fluids – internal boundaries within a single fluid can also transmit stress.

Because it is a surface force, stress acts only in a thin layer adjacent to the boundary, and the total force acting is proportional to the surface area of the plane that describes the boundary. The forces of this type that are of interest are pressure and friction. Pressure is a 'normal stress', and friction is a 'shear stress'.

There is a third type of force, called a line force, because it acts along a line. Surface tension is an example of a line force. Surface tension acts at fluid interfaces and so becomes important only when we consider boundaries such as that between the atmosphere and the ocean. In this book, we will neglect surface tension.

Let us consider each of the above forces in turn. First, we will consider the forces that are acting on a fluid regardless of the motion of the frame of reference. Recall that a non-accelerating frame of reference is called an inertial frame.

4.3 Forces in an inertial reference frame

4.3.1 Gravity

The force due to gravity originates from the mutual attraction between any object or fluid parcel and the Earth. Detailed observations of periodic motions in the solar system confirm the law, developed by Newton and published in his *Principia* in 1687, that every particle in the Universe attracts every other particle with a force proportional to the product of the masses and inversely proportional to the square of the distance between them. The universal constant of proportionality is $G = 6.673 \times 10^{-11} \, \text{N m}^2 \, \text{kg}^{-2}$. Strictly speaking, the inverse square law, which is expressed in Table 4.1, holds only for point masses, but for spheres of uniform density it is correct to consider the mass, however great, to be concentrated at the

Table 4.1 Forces acting on a fluid in an inertial frame

Force	Mathematical expression
Gravity	$\vec{g}^* = -\left(\dfrac{GM}{r^2}\right)\left(\dfrac{\vec{r}}{r}\right)$
Pressure gradient force	$\vec{P}_g = -\dfrac{1}{\rho}\left(\dfrac{\partial p}{\partial x}\vec{i} + \dfrac{\partial p}{\partial y}\vec{j} + \dfrac{\partial p}{\partial z}\vec{k}\right)$
Viscous force	$\vec{F}_r = \nu\left[\left(\dfrac{\partial^2 u}{\partial x^2} + \dfrac{\partial^2 u}{\partial y^2} + \dfrac{\partial^2 u}{\partial z^2}\right)\vec{i} + \left(\dfrac{\partial^2 v}{\partial x^2} + \dfrac{\partial^2 v}{\partial y^2} + \dfrac{\partial^2 v}{\partial z^2}\right)\vec{j} + \left(\dfrac{\partial^2 w}{\partial x^2} + d\dfrac{\partial^2 w}{\partial y^2} + \dfrac{\partial^2 w}{\partial z^2}\right)\vec{k}\right]$

center. Of course, the Earth is neither spherical nor of uniform density; nevertheless, the relationship holds to a sufficient level of accuracy for our purposes. Hence, by substituting for the mass and mean radius of the Earth (Appendix B), we arrive at a value of the gravitational force per unit mass acting at the surface of the Earth, g_0^*, of $9.83\,\mathrm{m\,s^{-2}}$.

Gravity is a conservative force; that is, by definition, the work done by gravity is independent of the path, and indeed the work done by gravity around a closed path is identically zero. We know from the discussion in Chapter 2 that this implies that we can write the gravity vector g^* as the gradient of a scalar potential function. In the case of gravity, this function is known as the *geopotential* and because we assume, for now, that the gravity vector has a component only in the z direction, this function varies only with z. The geopotential is designated by the symbol Φ:

$$\Phi(z) = \int_0^z g^*(z)\,dz \tag{4.2}$$

Though small, the variations in gravity can be accounted for by using the geopotential. Since a surface of constant geopotential represents a surface along which all objects of the same mass have the same potential energy, if gravity were constant, a geopotential surface would also have the same altitude everywhere. Since gravity is not constant, a geopotential surface will have varying altitude. The *geopotential height* is defined as

$$Z = \frac{\Phi}{g_0^*} = \frac{1}{g_0^*}\int_0^z g^*\,dz \tag{4.3}$$

Because the variations in gravity are small, the geopotential height is very close to the actual (geometric) height, particularly close to the surface. Hence, in many situations we can use a constant value of g and the geometric height. Nevertheless, it is important to be aware of the approximation.

4.3.2 Pressure gradient force

A pressure *force* can only be exerted within a fluid if there is a difference in pressure from one parcel to another; otherwise the parcels are exerting an equal and opposite force on each other, and there is no net force. Therefore, the force associated with pressure arises from the *pressure gradient*, not the pressure itself.

To derive the mathematical form of the pressure gradient force, we consider an infinitesimal parcel of air in the presence of a pressure gradient (Figure 4.1). The force due to atmospheric pressure acting on the left hand face of the parcel is

$$pA = p\delta y \delta z$$

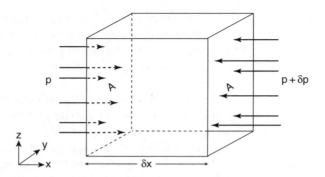

Figure 4.1 The x component of the pressure gradient forcing acting over an area A on the surface of an infinitesimal parcel of air

and on the right hand face is simply

$$-(p+\delta p)\,\delta y \delta z = -\left(p + \frac{\partial p}{\partial x}\delta x\right)\delta y \delta z$$

if the dimensions of the parcel are small compared to the scale of the pressure changes. Hence, the x component of the net force exerted by this pressure field on the parcel is simply the sum of the forces acting on either side of the parcel:

$$-\frac{\partial p}{\partial x}\delta x \delta y \delta z$$

In order to write this in terms of a force per unit mass, we must divide the expression by the mass, $\rho \delta x \delta y \delta z$, to get

$$\vec{P}_g \bullet \vec{i} = -\frac{1}{\rho}\frac{\partial p}{\partial x} \qquad (4.4)$$

The negative sign shows that the force is directed from high pressure to low pressure, as we would expect. The other components of the force as shown in Table 4.1 can be derived in an analogous manner.

Example The pressure gradient force at any location can be easily calculated using station reports or a mean sea level pressure map. Consider our case study storm at 12 UTC on Friday, 14 February 2003 (Figure 4.2). The low is situated over western Kansas at this time and there is a sea level pressure report of 1000.3 hPa from Dodge City, near the center of the low. The temperature is reported to be 48°F (9°C) at Dodge City. Around 500 km away, in Limon, Colorado, the sea level pressure is

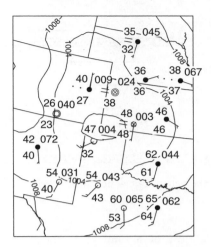

Figure 4.2 Sea level pressure and station reports over Oklahoma and surrounding states on 14 February 2003 at 12 UTC

reported to be 1000.9 hPa. From this information and using the ideal gas equation, we can calculate the pressure gradient force acting:

$$\rho = \frac{p}{R_d T} \approx \frac{100\,030}{287 \times (9+273)}$$

$$\approx 1.2 \text{ kg m}^{-3}$$

$$\vec{P}_g = -\frac{1}{\rho}\frac{\partial p}{\partial n}\vec{n}$$

$$\approx -\frac{1}{1.2}\frac{100\,090 - 100\,030}{500 \times 10^3}\vec{n}$$

$$\approx -9.7 \times 10^{-5}\vec{n}$$

where we have assumed a coordinate system with unit vector $\vec{n}$ directed outward from the center of the low, for simplicity. The resulting pressure gradient force of 9.7×10^{-5} N kg^{-1} is then directed toward the center of the low, as expected. We have also allowed the density at Dodge City to represent the density of air in the entire region – this will introduce a small error.

4.3.3 Viscous force

The effect of viscosity, or friction, within a fluid, results in a surface force like pressure, and its basis is expressed in the same way, namely as a force per unit area. In order to illustrate how the viscous force behaves, we can consider a classic

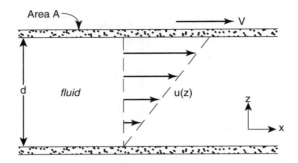

Figure 4.3 Flow of a fluid through a channel bounded by two plates, one moving with speed V (top), and one stationary

example: the steady state flow of a fluid in a channel between two solid plates (Figure 4.3). The upper plate is moving at some slow speed V and the lower plate is stationary. There is no variation of the resulting flow with time or in the x or y directions.

Recall that in Chapter 3 we characterized a fluid as a material that yields continuously to any shear stress, no matter how small. Although it is not self-evident, from empirical evidence it turns out that the relative velocity of a fluid is exactly zero at the boundary with a solid. That is, the fluid molecules adjacent to the solid move with the same velocity as the solid. This is because the solid surface is exerting a *shear stress* on the fluid. In the example of the channel, this means that the fluid adjacent to the upper plate is moving at speed V and the fluid adjacent to the lower plate is stationary. Measurements from this experiment show that the magnitude of the shear stress $\tau = F/A$ exerted on the fluid by the upper plate is proportional to the speed of the upper plate and inversely proportional to the distance d between the plates:

$$\frac{F}{A} \propto \frac{V}{d}$$

The constant of proportionality can also be measured, and from such measurements it is known that this constant, designated the coefficient of shear viscosity μ, is dependent upon the fluid in question. That is, this coefficient depends on the molecular properties of the fluid constituents. This relationship holds regardless of the distance d between the plates, and hence we can write the x component of the shear stress acting on the uppermost fluid layer in the channel as

$$\vec{\tau} \bullet \vec{i} = \mu \frac{\partial u}{\partial z} \tag{4.5}$$

The moving molecules in the fluid exert a shear stress on the layer of fluid below them, causing them to move also, but this layer is also being acted upon by the layer below that, exerting a retarding stress, and so on, down through the layers, creating a linear profile of fluid flow as shown in Figure 4.3.

We have defined our example to be steady state; that is, although there is motion, there is *no acceleration*. Hence, there can be *no net viscous force* acting. Thus, just as in the case of pressure, the mere presence of shear stress (internal friction) in a fluid does not cause a net force. In order for this to occur, there must be a *gradient* in the shear stress. This may be derived in an analogous manner to the derivation of the pressure gradient force in the previous section, resulting in an expression of the form

$$\frac{1}{\rho}\frac{\partial \tau}{\partial z} = \frac{\mu}{\rho}\frac{\partial^2 u}{\partial z^2}$$

but with one important distinction. The shear stress must be considered to be a vector quantity, rather than a scalar like pressure, and hence a more general expression of the x component of the viscous force (or 'shear stress gradient force') in a truly three-dimensional example is

$$\vec{F}_r \bullet \vec{i} = \frac{\mu}{\rho}\left(\frac{\partial^2 u}{\partial x^2} + \frac{\partial^2 u}{\partial y^2} + \frac{\partial^2 u}{\partial z^2}\right) \quad (4.6)$$

We can replace the coefficient of shear viscosity in Equation (4.6) by a new quantity

$$\nu = \frac{\mu}{\rho} \quad (4.7)$$

which is known as the *kinematic viscosity coefficient*. For standard conditions at sea level the kinematic viscosity of the atmosphere is around $1.5 \times 10^{-5}\,\mathrm{m^2\,s^{-1}}$, and of the ocean is around $1.0 \times 10^{-6}\,\mathrm{m^2\,s^{-1}}$.

4.3.4 Hydrostatic balance

Consider the atmosphere at rest. There is a downward force acting at all times, which is simply the weight of the air under the influence of the gravitational force. For the atmosphere to be at rest, there can be no net force present and hence there must be a balancing upward force. This is provided by the pressure gradient force in the vertical, directed from higher pressure near the surface to lower pressure aloft (see Figure 1.2). This state of balance between the two forces is known as *hydrostatic balance*, and is expressed mathematically as

$$-\frac{1}{\rho}\frac{dp}{dz} = g^*$$

$$\Rightarrow \frac{dp}{dz} = -\rho g^* \quad (4.8)$$

Note that we do not use the partial derivative of pressure in this expression, since it is only the variations of pressure with height that are permitted (horizontal variations could give rise to motion). This equilibrium is in fact a good approximation even when the atmosphere is in motion, because, as we will see, motions that cause departures from this balance tend to be comparatively small.

In practice, the measurement of density in the atmosphere is more difficult than the measurement of pressure or temperature. We can eliminate this quantity by combining Equation (4.8) with the ideal gas equation (Equation (3.1)) to give

$$dp = -\rho g^* dz$$

$$= -\frac{p}{R_d T} g^* dz$$

$$\frac{dp}{p} = -\frac{g^*}{R_d T} dz$$

$$\ln\left(\frac{p}{p_0}\right) = -\int_0^z \frac{g^*}{R_d T} dz \tag{4.9}$$

If we assume that the integrand is approximately a constant, by using a layer average temperature (denoted by angle brackets below), and we define that integrand to be a quantity known as the *scale height*,

$$H = \frac{R_d \langle T \rangle}{g^*} \tag{4.10}$$

we can write

$$p \approx p_0 e^{-z/H} \tag{4.11}$$

Hence, the altitude is proportional to the logarithm of the pressure. Equation (4.11) is one form of what is known as the *hypsometric equation*, which relates the pressure and altitude. In the Earth's atmosphere, the scale height, at which the pressure drops by a factor of $1/e$, is around 8 km.

In Section 4.3.1 we defined the geopotential, allowing us to take account of the variation in g. So if we assume instead that the integrand in Equation (4.9) is not constant, we can write

$$\frac{dp}{p} = -\frac{g^*}{R_d T} dz$$

$$d\Phi = -R_d T \frac{dp}{p}$$

$$\Delta\Phi = -R_d \int_{p_0}^{p} T d\ln p$$

$$\Rightarrow \Delta Z = -\frac{R_d}{g_0^*} \int_{p_0}^{p} T d\ln p \tag{4.12}$$

The difference in geopotential height between two pressure levels derived here is also called the *thickness* of the layer, and is proportional to the mean temperature in the layer. Thus, we can see that as a layer of air warms, it expands and its thickness increases.

Example Figure 4.4 shows a vertical profile of temperature and dew point temperature at Upton, NY on 00 UTC 16 Feb 2003. This data is plotted on a skew T thermodynamic diagram. On skew T diagrams the temperature axis is 'skewed' such that lines of constant temperature slope from the lower left to the upper right of the diagram. The vertical axis on the diagram is logarithmic in pressure, with pressure decreasing from the bottom to the top of the diagram. The uses of the diagram are explored further in Chapter 11.

Using this diagram we can determine the thickness of the layer between 500 and 400 hPa, using Equation (4.12) to calculate ΔZ.

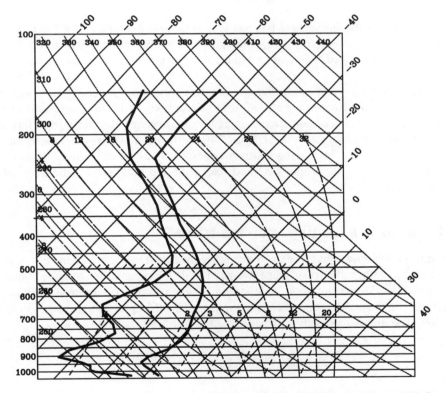

Figure 4.4 'Skew T' atmospheric sounding from 00 UTC 16 Feb 2003 at Upton, NY showing temperature (right curve) and dew point temperature (left curve) expressed in °C against pressure (hPa). On this diagram pressure is indicated by the vertical scale on the left side of the diagram. Lines of constant temperature are 'skewed' and slope from the lower left to the upper right and are labeled along the top and right sides of the diagram

Using the layer-averaged temperature will simplify the calculation. From Figure 4.4, the temperature at 500 hPa is $-20°C$ and the temperature at 400 hPa is $-31°C$. This gives an average temperature (since the temperature variation is assumed to be roughly linear throughout this layer) of $-25.5°C$. Converting the temperature from °C to K and using SI units gives

$$\Delta Z = -\frac{R_d}{g_0^*} \int_{p_0}^{p} T d\ln p$$

$$= \frac{-287}{9.85} \int_{50\,000}^{40\,000} 247.5 d\ln p$$

$$= \frac{-287}{9.85} \times 247.5 \times \ln \frac{40\,000}{50\,000}$$

$$= 1609 \text{ m}$$

Had this layer of the atmosphere had an average temperature of 277.5 K (30 K warmer than was actually observed) the layer thickness would have been 1804 m. Conversely a colder layer-averaged temperature would have resulted in a smaller thickness.

Other items to note on this sounding are that the temperature generally decreases with decreasing pressure (increasing altitude). The exceptions to this are the layers between 950 and 800 hPa where temperature increases with increasing altitude (referred to as an *inversion*) and in the layer from 240 to 150 hPa where the temperature is nearly constant with increasing altitude (referred to as an *isothermal layer*). The base of this layer is the tropopause and marks the top boundary of the troposphere (Section 1.2).

4.4 Forces in a rotating reference frame

The Earth is rotating about its axis. Since it is convenient to adopt a frame of reference fixed to the Earth, rather than space, we need to develop equations of motion that are appropriate for a rotating coordinate system. Such a frame of reference is known as a non-inertial reference frame because the frame itself imparts an acceleration to the motion. The angular velocity of rotation of the Earth, Ω, is small:

$$\Omega = \frac{2\pi}{1 \text{ day}} = \frac{2\pi}{86\,400 \text{ s}} = 7.292 \times 10^{-5} \text{ s}^{-1}$$

and hence for many phenomena, the effects of the rotation of the reference frame are negligible. However, on the space and time scales of some atmospheric motions, the effect is important and must be included. There are two aspects to this effect, which we will deal with in turn: forces on an object or parcel at rest, and forces on a moving parcel.

4.4.1 Centrifugal force

Consider a student standing at the center of a table that is rotating with angular velocity ω in a laboratory (Figure 4.5). He conducts an experiment in which he places an object of mass m at some distance r from him on the table. He finds that, if the forces of friction between the object and the table are negligible, the object moves radially away from him. Then, the student attaches a string of length r to the object, and measures the force required to hold the object in place. Using a series of measurements using different values of m and r, the student reports that the force required to hold the object in place acts radially toward to him and can be described by the expression $mr\omega^2$. Since the object is at rest from the point of view of the student, he concludes that, in order for there to be no net force, there must be an equal and opposite force acting radially outward. This outward radial force is called the *centrifugal force*.

Meanwhile, a second student is observing the experiment while standing on the floor beside the table. Her observations are quite different. She notes that the object, while being held by the string, is in fact accelerating because it is rotating at the same rate as the table. Hence, in her frame of reference, there is a net force acting on the object, exerted by the student on the table. This force acting radially inward from the object is called the *centripetal force* and can be described by the expression $mr\omega^2$.

Hence, the centrifugal force arises only if the observation is taken in the rotating frame of reference, and derives directly from the acceleration of the frame. This is why the centrifugal force is often called an apparent force. However, to take account of the rotating frame of reference, this force must be treated as any other force acting on an object.

If the Earth were a smooth sphere, then an object placed at any point on the Earth would have, as in the case of the rotating table, a tendency to move away from the center of rotation, as shown in Figure 4.6(a). This means that there would be a

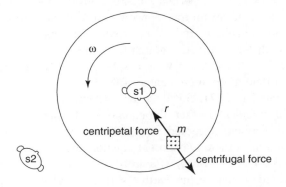

Figure 4.5 Laboratory experiment in which two students (s1 and s2) observe the motion of an object held on a string by s1 on a spinning frictionless table

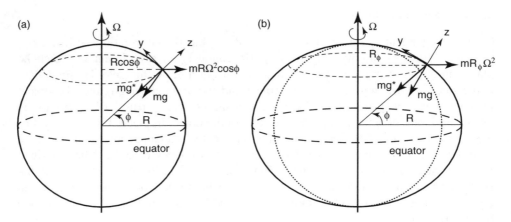

Figure 4.6 Centrifugal and gravitational forces acting on an object of mass m at rest at latitude ϕ on (a) a perfectly spherical Earth and (b) the actual Earth (not to scale). Note that the centrifugal force is in fact almost three orders of magnitude smaller than the gravitational force

component of the centrifugal force in the positive z direction, and a component in the negative y or equatorward direction.

The centrifugal force is a body force; that is, it acts through the center of mass of the object. Hence, it can be combined with the gravity force to create a composite body force. This results in a quantity known as the *effective gravity*, whose value at sea level is 9.81 m s^{-2}, slightly reduced from the value calculated in Section 4.3.1 because of the centrifugal force component in the positive z direction. The component of the centrifugal force in the negative y direction results in a change in the direction of the effective gravity force relative to the gravity force, and thus the effective gravity force is not directed at the center of the Earth. The effective gravity is denoted g, and can replace g^* in all of the previous equations in this chapter, including Equation (4.8) above, which expresses the hydrostatic balance.

However, the Earth is not a perfect sphere. Consider the distance between two points on the equator that are on the 'opposite sides of the Earth' from each other: Quito, Ecuador and Pekanbaru, Indonesia provide an approximate example. If we could measure this distance directly though the center of the Earth, we would find that it is $12\,756\,272$ m. If the Earth were a sphere, the distance between the North Pole and the South Pole through the center of the Earth would also be $12\,756\,272$ m, but in fact this distance is $12\,713\,504$ m, around 43 km or 0.34% smaller. The reason for this 'equatorial bulge' is, in fact, the equatorward component of the centrifugal force: as the Earth formed, it cooled from a liquid state as it spun, and the mass distribution adjusted until there was no net equatorward component in the net force represented by the effective gravity (Figure 4.6b). The result is that the effective gravity is everywhere normal to the Earth's surface (neglecting topography) with a magnitude of 9.81 m s^{-2}, and objects at rest on the surface of the Earth experience no net equatorward force.

4.4.2 Coriolis force

The second influence of the rotating frame becomes evident when the object in question starts to move *relative* to the frame of reference; for example, a parcel of air moving on the rotating Earth. This results in apparent acceleration and an associated force, the Coriolis force, which is named after the nineteenth-century French mathematician Gaspard Gustave de Coriolis. There are two aspects to the Coriolis force. The first is due to an additional centrifugal force that arises from the relative motion. The second is due to changes in the relative angular momentum of the object. We will now look at each of these aspects of the Coriolis force.

First, consider a situation where an object (such as a parcel of air) located at some latitude and longitude is started impulsively from rest in the eastward direction. Assuming the absence of friction and the pressure gradient force (apart from that imparting hydrostatic balance in the vertical), the object subsequently experiences no forces, and so moves with a constant zonal velocity $\vec{u} = (u, 0, 0)$. However, now its rate of rotation has increased relative to that of the Earth, from Ω to $\Omega + u/R_\phi$. This increases the centrifugal force which appears to act in this frame of reference. This alters the balance between the gravitational force and the centrifugal force, and so from the frame of reference of the Earth, we will observe an additional, radially outward, force.

We can express the total centrifugal force mathematically as follows:

$$CF_{total} = m\left(\Omega + \frac{u}{R_\phi}\right)^2 \vec{R}_\phi$$

$$= m\Omega^2 \vec{R}_\phi + \frac{2m\Omega u}{R_\phi}\vec{R}_\phi + m\frac{u^2}{R_\phi^2}\vec{R}_\phi$$

where the notation is consistent with Figure 4.6(b). Since $R_\phi \gg u$ for the wind speeds we typical experience on the Earth, $u^2/R_\phi^2 \ll 1$, and we can neglect the final term in the expression. The first term is simply the centrifugal force for a parcel at rest, and can be incorporated into the gravity term to yield the effective gravity as before. This leaves a remaining component of centrifugal force which is due to the motion of the object. We can decompose this into components based on the geometry shown in Figure 4.6, to give

$$CF_{motion} = -2m\Omega u \sin\phi \vec{j} + 2m\Omega u \cos\phi \vec{k} \qquad (4.13)$$

So, in the inertial reference frame, an observer would conclude that the centripetal force provided by effective gravity was not in balance with the vertical component of the pressure gradient force, and would lead to additional relative motion of the object causing the object to move upward and equatorward.

However, this centrifugal force is only observed while on the rotating frame in the presence of zonal motion. Meridional and vertical motion relative to the surface of the Earth do not give rise to an additional component of the centrifugal force, but these motions also give rise to apparent forces on our rotating frame of reference on the Earth. We will now look at this response to meridional and vertical motion.

Once again, consider an object which is moving at a constant northward velocity, subject to no net forces. At a given moment, the object passes through a latitude circle ϕ, and possesses an angular momentum given by

$$I\omega = \tfrac{1}{2} m R_\phi^2 \Omega$$

where I is the moment of inertia of the object. As the object continues northward, it will conserve its angular momentum even as it moves to a new latitude circle which, being further north, is a smaller distance from the center of rotation. We can write this conservation equation as

$$\tfrac{1}{2} m R_\phi^2 \Omega = \tfrac{1}{2} m \left(R_\phi + \delta R \right)^2 \left(\Omega + \frac{\delta u}{R_\phi + \delta R} \right)$$

$$R_\phi^2 \Omega = R_\phi^2 \Omega + 2 R_\phi \delta R \Omega + \frac{R_\phi^2 \delta u}{R_\phi + \delta R} \tag{4.14}$$

where we have neglected second-order terms, and made the assumption that the angular momentum will be balanced at the new location by an increase in the angular velocity of the object. Solving this equation for δu, further neglecting second-order terms, and noting that $\delta R = -v \sin \phi \, \delta t$, we find

$$\delta u = 2 \Omega v \sin \phi \, \delta t$$

Taking the limit as $\delta t \to 0$ we can write this as

$$\frac{du}{dt} = 2 \Omega v \sin \phi \tag{4.15}$$

Hence, the requirement for the conservation of angular momentum results in a zonal acceleration of the object which is proportional to the meridional speed of the object.

Now, consider the conservation of angular momentum for an object which is moving with a constant vertical velocity. In this case, the latitude is not changing, but the distance to the center of rotation is changing due to the vertical motion. In this case, we again solve for δu in Equation (4.14) but note that $\delta R = w \cos \phi \, \delta t$. Taking the limit as before, we find that

$$\frac{du}{dt} = -2 \Omega w \cos \phi \tag{4.16}$$

Finally, we can note that for an object moving with a constant zonal velocity there is no change in angular momentum, since both the angular velocity of rotation, Ω, and the distance to the center of rotation, R, are not changing.

Hence, we have derived, using two different perspectives, the acceleration that is apparent due to relative motion in a rotating frame of reference in the zonal, meridional, and vertical directions. All of these accelerations are typically grouped in a single apparent vector force known as the Coriolis force, which can be written in total as

$$\frac{F_{Coriolis}}{m} = (2\Omega v \sin\phi - 2\Omega w \cos\phi)\vec{i} - 2\Omega u \sin\phi \vec{j} + 2\Omega u \cos\phi \vec{k} \quad (4.17)$$

We can, in fact, write this as the combination of two vectors, the wind vector and the Earth's rotation vector, which in components is $\vec{\Omega} = \Omega\cos\phi\vec{j} + \Omega\sin\phi\vec{k}$ (see Figure 4.6). Using this, we can write

$$\frac{F_{Coriolis}}{m} = -2\vec{\Omega} \times \vec{u} \quad (4.18)$$

Example The Coriolis force per unit mass at any location can be easily calculated using station reports. Consider our case study storm at 12 UTC on Friday 14 February 2003 (see Figure 4.2). In Limon, the wind report is 5 kts, from the north. The latitude at Limon is 39°44′N. From this, we find

$$\text{wind speed} = 5 \text{ kts} = 2.6 \text{ m s}^{-1}$$
$$u = 0 \quad v = -2.6$$

$$\phi = 39°44' = 39.73°$$
$$\Omega = 7.292 \times 10^{-5} \text{ s}^{-1}$$

$$\frac{F_{Coriolis}}{m} = (2\Omega v \sin\phi - 2\Omega w \cos\phi)\vec{i} - 2\Omega u \sin\phi \vec{j}$$
$$+ 2\Omega u \cos\phi \vec{k}$$
$$= 1.46 \times 10^{-4} \left[(-2.6 \times \sin 39.73)\vec{i} + 0\vec{j} + 0\vec{k} \right]$$
$$= -2.4 \times 10^{-4}\vec{i} + 0\vec{j} + 0\vec{k}$$

The Coriolis force per unit mass is directed toward the east; that is, to the right of the wind velocity. It is of a similar order of magnitude to the pressure gradient force of 9.7×10^{-5} N kg^{-1}, which is directed toward the center of the low (calculated for this location in Section 4.3.2).

In general, then, the Coriolis force does no work, but deflects the flow of air parcels to the right in the Northern Hemisphere and to the left in the Southern Hemisphere. This force cannot initiate motion.

4.5 The Navier–Stokes equations

Forces cause acceleration. When there are no forces, or when all the forces cancel each other, there is no acceleration. We have seen that the primary forces in the vertical direction are gravity and the centrifugal force acting downward and the pressure gradient force acting upward. There is also a small vertical component in the Coriolis force. The forces acting in the horizontal direction are the pressure gradient force, the viscous force, and the Coriolis force. We can use this information to construct an equation that calculates the net force acting on a parcel of air, and hence its acceleration. If we can perform this calculation at every point in the atmosphere, then, in theory, we can determine the winds anywhere on the Earth. To construct the equation, we simply sum all of the forces acting in each vector direction, and equate this sum to the acceleration of the parcel. Note that forces affect fluid parcels, and hence the acceleration is represented by the material derivative of wind with respect to time (see Section 2.8):

$$\frac{Du}{Dt} = -\frac{1}{\rho}\frac{\partial p}{\partial x} + \nu\left(\frac{\partial^2 u}{\partial x^2} + \frac{\partial^2 u}{\partial y^2} + \frac{\partial^2 u}{\partial z^2}\right) + (2\Omega v \sin\phi - 2\Omega w \cos\phi)$$

$$\frac{Dv}{Dt} = -\frac{1}{\rho}\frac{\partial p}{\partial y} + \nu\left(\frac{\partial^2 v}{\partial x^2} + \frac{\partial^2 v}{\partial y^2} + \frac{\partial^2 v}{\partial z^2}\right) - (2\Omega u \sin\phi) \qquad (4.19)$$

$$\frac{Dw}{Dt} = -g - \frac{1}{\rho}\frac{\partial p}{\partial z} + \nu\left(\frac{\partial^2 w}{\partial x^2} + \frac{\partial^2 w}{\partial y^2} + \frac{\partial^2 w}{\partial z^2}\right) + (2\Omega u \cos\phi)$$

↑ acceleration of air parcel   ↖ gravity   ↖ pressure gradient force   ↑ viscous force   ↑ Coriolis force

This set of equations is usually called the Navier–Stokes equations for the conservation of momentum. The equations are named after Claude Louis Navier (1785–1836) and George Gabriel Stokes (1819–1903), who both contributed to the development of the equations. Navier developed the form of these equations for an *incompressible* (that is, constant density) fluid in 1821. In 1822, he published a further refinement for viscous fluids, despite not having access to a physical theory for shear stress in a fluid. Nevertheless, he arrived at the proper form for these equations. Twenty years later, Stokes published papers on incompressible fluid motion, and then continued his investigations by addressing the problem of internal friction in fluids in motion from a more theoretical perspective.

We can write the set of equations in some alternative forms. First, since we expect to be solving these equations in an Eulerian framework, it is often appropriate to replace the material derivative with the local derivative. Based on the derivation in Section 2.8, then, we can write

$$\frac{\partial u}{\partial t} + u\frac{\partial u}{\partial x} + v\frac{\partial u}{\partial y} + w\frac{\partial u}{\partial z} = -\frac{1}{\rho}\frac{\partial p}{\partial x} + \nu\left(\frac{\partial^2 u}{\partial x^2} + \frac{\partial^2 u}{\partial y^2} + \frac{\partial^2 u}{\partial z^2}\right)$$
$$+ (2\Omega v \sin\phi - 2\Omega w \cos\phi)$$

$$\frac{\partial v}{\partial t} + u\frac{\partial v}{\partial x} + v\frac{\partial v}{\partial y} + w\frac{\partial v}{\partial z} = -\frac{1}{\rho}\frac{\partial p}{\partial y} + \nu\left(\frac{\partial^2 v}{\partial x^2} + \frac{\partial^2 v}{\partial y^2} + \frac{\partial^2 v}{\partial z^2}\right) - (2\Omega u \sin\phi)$$

$$\frac{\partial w}{\partial t} + u\frac{\partial w}{\partial x} + v\frac{\partial w}{\partial y} + w\frac{\partial w}{\partial z} = -g - \frac{1}{\rho}\frac{\partial p}{\partial z} + \nu\left(\frac{\partial^2 w}{\partial x^2} + \frac{\partial^2 w}{\partial y^2} + \frac{\partial^2 w}{\partial z^2}\right) + (2\Omega u \cos\phi)$$

Clearly, this is a very long-winded way of writing the equation set, and hence typically the material derivative form is written.

In addition, we can write this set of equations as a single vector equation that solves for $D\vec{u}/Dt$, rather than three components. Such notation is much more efficient, and allows the interpretation of the physical terms in the equation in a more straightforward way. However, for the full Navier–Stokes equations this form requires the use of vector calculus operators, which is beyond the scope of this book. Hence, in this case we will retain the component equations form.

4.5.1 Perturbation pressure

The variations in pressure that give rise to the pressure gradient force are quite small compared to the background pressure field that is in balance with the effective gravity force field. From the perspective of atmospheric motions, it is the pressure gradient that is of primary importance. Hence, one way to alter the perspective provided by the Navier–Stokes equations is to subtract a reference pressure field from the pressure gradient force term.

Recall that the pressure distribution in the vertical for the atmosphere at rest can be represented in part by the hydrostatic pressure, denoted p_0:

$$\frac{dp_0}{dz} = -\rho_0 g$$

For this application, we have also defined a corresponding reference density ρ_0. This hydrostatic pressure is a function only of height, and so we can define a new quantity by

$$p = p_0(z) + p_d$$

where p_d is called the perturbation pressure or dynamic pressure. It is generally true in the study of atmospheric motion that most of the vertical pressure gradient is balanced hydrostatically, and vertical motions result from very small departures from that balance.

Using this definition in the z component of the Navier–Stokes equations, we get

$$\rho\frac{Dw}{Dt} = -\rho g - \frac{\partial}{\partial z}(p_0(z)+p_d) + \rho\nu\left(\frac{\partial^2 w}{\partial x^2}+\frac{\partial^2 w}{\partial y^2}+\frac{\partial^2 w}{\partial z^2}\right)+\rho(2\Omega u\cos\phi)$$

$$= -\rho g + \rho_0 g - \frac{\partial p_d}{\partial z} + \rho\nu\left(\frac{\partial^2 w}{\partial x^2}+\frac{\partial^2 w}{\partial y^2}+\frac{\partial^2 w}{\partial z^2}\right)+\rho(2\Omega u\cos\phi)$$

$$\frac{Dw}{Dt} = -\frac{(\rho-\rho_0)}{\rho}g - \frac{1}{\rho}\frac{\partial p_d}{\partial z}+\nu\left(\frac{\partial^2 w}{\partial x^2}+\frac{\partial^2 w}{\partial y^2}+\frac{\partial^2 w}{\partial z^2}\right)+(2\Omega u\cos\phi)$$

where now the equation is in terms of the perturbation pressure, and the effective gravitational force is replaced by a buoyancy force.

It is important to note that the definition of reference pressure and density is not unique. Hence, while the total force acting on a parcel of air is independent of the form of the equation, the partition of this unique force field between the perturbation pressure gradient and the buoyancy force is not unique.

For example, using the background hydrostatic pressure profile of the environment as our reference pressure and density, we would conclude that the updrafts in severe thunderstorms are negatively buoyant at the cloud base (that is, they cause sinking motion) and do not become positively buoyant until some kilometers higher in the cloud. In this perspective, the updraft is actually driven by a large perturbation pressure gradient that compensates for the negative buoyancy. However, if we used the thunderstorm rather than the environment as our reference profile for pressure and density, we would conclude a forcing that consists of zero or positive buoyancy and a smaller perturbation pressure gradient.

Review questions

4.1 Calculate the magnitude of $\vec{g}^*$ at:

(a) sea level at the equator and at the North Pole;

(b) at the equator at altitudes of 0, 10, 100, and 1000 km above sea level.

4.2 Using the surface weather map shown in Figure 1.7 calculate the magnitude of the pressure gradient force between

(a) Minneapolis, Minnesota and Des Moines, Iowa

(b) Charleston, South Carolina and Atlanta, Georgia.

(c) How does the spacing of the sea level pressure isobars differ between these cities and how does this spacing relate to the magnitude of the pressure gradient force calculated for parts (a) and (b) of this question?

4.3 As a thunderstorm passes over an area, typically the air temperature will decrease and the surface pressure will increase. The increase in pressure is a hydrostatic response to the changing temperature (and density) of the air. On 6 May 2005 a weather station in Boulder, Colorado reported a decrease in temperature from 74°F to 59°F (23°C to 15°C) and an increase in pressure from 826 hPa to 827.3 hPa as a thunderstorm passed over the station. Assume that the cooling of the air was uniform over some depth of the atmosphere. Calculate the depth of air that would need to be cooled by the observed amount to produce the observed change in pressure.

4.4 (a) Write an equation for the geopotential height at a given pressure level for an atmosphere in which the vertical temperature profile is given by $T(z) = T_0 + \Gamma z$, where T_0 is the temperature at an elevation of 0 m and $\Gamma = -dT/dz$ is the *lapse rate*.

(b) Calculate the height of the 300 hPa pressure surface for $\Gamma = 6°C\,km^{-1}$ and $T_0 = -30, 0,$ and $30°C$.

4.5 At 00 UTC 15 Feb 2003, the radiosonde data from Davenport, Iowa indicated a temperature of $-2.9°C$ at 850 hPa and $-16.3°C$ at 500 hPa. At the same time the radiosonde at Little Rock, Arkansas indicated a temperature of $10.4°C$ at 850 hPa and $-14.5°C$ at 500 hPa. Using this data, calculate the thickness of the 850 to 500 hPa layer at Davenport and at Little Rock. Discuss the difference in thickness of this layer at these two cities, based on the information shown on the surface weather map in Figure 1.7.

4.6 Isolines of 1000 to 500 hPa thickness are often drawn on surface weather maps using a contour interval of 60 m. What is the corresponding layer mean temperature interval?

4.7 Consider the sounding shown in Figure 4.7, taken at McGrath, Alaska at 12 UTC 26 Aug 2002.

(a) Which curve is dew point temperature and which curve is temperature? Why?

(b) There are two significant temperature inversions. At what levels do they begin?

(c) Calculate the thickness of the 600 to 300 hPa layer in this sounding.

4.8 Show that Equations (4.17) and (4.18) are equivalent.

4.9 Consider the map of sea level pressure shown in Figure 4.8, which shows a low-pressure system that occurred in August 2000 in the Chukchi Sea north of Alaska. This system caused record-breaking high winds along the Alaskan north coast.

(a) Calculate the total horizontal pressure gradient force between the center of the low and the center of the high-pressure system to its west, assuming an air density of $1\,kg\,m^{-3}$. At this latitude, one degree of longitude is approximately equivalent to 28 km.

FUNDAMENTAL FORCES

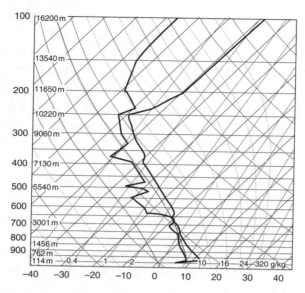

Figure 4.7 Radiosonde sounding from McGrath, AK on 12 UTC 22 Aug 2002

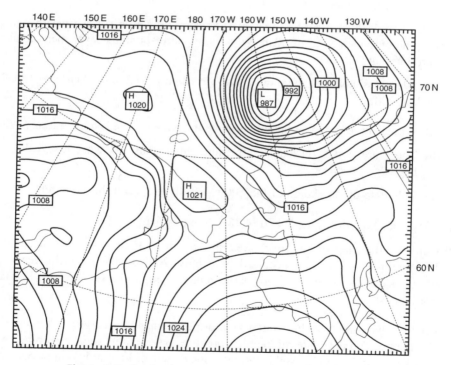

Figure 4.8 Sea level pressure over Alaska during August 2000

(b) A 1 kg parcel of air starts from rest at 175°W, 75°N. In what direction would it be accelerated if it was under the influence alone of the above pressure field? How many hours does it take for the parcel to achieve a speed of 35 m s^{-1}?

(c) Suppose the high-pressure system was displaced 50 km further to the west. Without performing an additional calculation, would the parcel take more or less time to reach this speed? Why?

4.10 (a) Use the surface weather observations from 00 UTC 17 Feb 2003 (Figure 1.11) to calculate the horizontal components of the pressure gradient force and Coriolis force at Davenport, Iowa. (Hint: use the sea level pressure observations from Chicago, Illinois (ORD), St. Louis, Missouri (STL), Des Moines, Iowa (DSM), and Madison, Wisconsin (MSN) to calculate the pressure gradient force.)

(b) What additional force would be required to have no net acceleration of the air at Davenport?

4.11 (a) While skiing in Colorado, at a latitude of 40°N, I become airborne after jumping off the lip at the top of a ski run at a Colorado resort that faces north. I commence my flight at 10 m s^{-1} traveling parallel to the 42° slope, and make contact with the snow 50 m down the slope. Calculate in centimeters to three significant figures my displacement due to the Coriolis force by the time I touch the surface. Assume that the Coriolis force remains constant during the flight at the initial value at lift-off.

(b) Without recalculating, what is the displacement if the speed is 20 m s^{-1}? Is the Coriolis force larger or smaller? Why?

5 Scale analysis

5.1 Dimensional homogeneity

Consider the following empirical relationship, used as a rule of thumb by ocean weather forecasters to estimate the speed of travel of ocean waves, C_p, as a function of the wavelength (the distance between wave crests), l:

$$C_p = 1.34\sqrt{l} \qquad (5.1)$$

where l is measured in feet to yield a wave speed in knots. For this relationship to be useful as a forecasting tool it is crucial to know the requirements for the units, particularly in this case where they are not conventional SI units. A better formulation would be to write

$$C_p = \beta\sqrt{l} \quad \text{where} \quad \beta = 1.34 \frac{\text{knots}}{\text{feet}^{1/2}} \qquad (5.2)$$

or in conventional SI units

$$C_p = \beta\sqrt{l} \quad \text{where} \quad \beta = 0.38\,\text{m}^{1/2}\,\text{s}^{-1} \qquad (5.3)$$

The form of the relationship expressed in Equations (5.2) and (5.3) has the advantage that it can be tested easily for the property of *dimensional homogeneity*; that is, that all terms in an equation have the same dimensions. Such a property is a prerequisite for any physical equation, so that the numerical equality does not depend on the units of measurement as long as appropriate unit conversions are made.

In our example, we can see that wave speed has the dimensions, regardless of choice of units, of length $[L]$ divided by time $[T]$, which can be denoted $[LT^{-1}]$. In Equation (5.1), if we initially assume that the number 1.34 has no units, then the dimensions of the right hand side of the equation are simply $\left[L^{1/2}\right]$, which is clearly an unphysical result. If, on the other hand, we consider Equation (5.2) or (5.3), the term on the right hand side has the dimensions

$$[L]^{1/2}[T]^{-1}[L]^{1/2} = [LT^{-1}]$$

and so the equation is dimensionally homogeneous. This is a basic check which can be applied to any equation, but is especially useful in the analysis of complex equations that have no analytic solutions.

5.2 Scales

The atmosphere is a complex system capable of supporting many different types of motion from small turbulent eddies between two buildings to large weather systems. In order to make the theoretical study of a particular system easier, we can simplify our approach by knowing which scales are important in the driving mechanisms of that system. This allows us to determine whether we can ignore processes that are happening at different scales. Time scale is important as is length scale.

How do we determine these scales? For a particular type of system, we measure the magnitudes of the important variables, such as pressure, wind speed, and moisture mixing ratio. From these we determine the amplitudes and typical distances over which fluctuations occur. If we are looking at a wave-like phenomenon, we can measure variables like frequency and wavelength. From a series of measurements of many different examples of the same type of system, we can determine typical values for many of these quantities, and hence determine the appropriate scale. Some scales of some common phenomena are shown in Table 5.1.

Note that motion with a small spatial scale tends to have a short time scale, and vice versa. This is often, although not always, true.

5.3 Non-dimensional parameters

If the governing equation for a particular situation is known, the principal of dimensional homogeneity can be used to derive useful non-dimensional parameters. For

Table 5.1 Scales for some typical atmospheric phenomena. A planetary wave is a type of wave in the atmosphere that encircles the entire Earth (see Chapter 8). A foehn or a chinook is a strong wind that is created when air flows downhill from a high elevation (see Section 13.3) – such winds often have names associated with a particular mountain range

Type of motion	Horizontal length scale	Time scale
Cold front	50 km	1–2 days
Tornado	100 m	Minutes
Mid-latitude weather system	1000 km	Several days
Cumulus cloud	1 km	Tens of minutes
Surf	10 m	Seconds
Planetary wave	10 000 km	Weeks–months
European Alps foehn or Rocky Mountain chinook	10 km	Hours

example, consider the x component of the Navier–Stokes equations, which are applicable to any general atmospheric flow:

$$\frac{\partial u}{\partial t} + u\frac{\partial u}{\partial x} + v\frac{\partial u}{\partial y} + w\frac{\partial u}{\partial z} = -\frac{1}{\rho}\frac{\partial p_d}{\partial x} + \nu\left(\frac{\partial^2 u}{\partial x^2} + \frac{\partial^2 u}{\partial y^2} + \frac{\partial^2 u}{\partial z^2}\right)$$
$$+ (2\Omega v \sin\phi - 2\Omega w \cos\phi)$$

Since this physical equation is dimensionally homogeneous, all of the terms will have the same dimensions, which we can derive most easily from the first term:

$$\frac{[U]}{[T]} = \frac{[LT^{-1}]}{[T]} = \left[\frac{L}{T^2}\right]$$

where $[U]$ is the dimension for wind speed. Thus, we can expect that all terms in the equation will have the dimension $[L/T^2]$. A corollary of this is that the ratio of any two terms in the equation will be a non-dimensional number. Further, the magnitude of this number, based on the scales of the system of interest, can tell us something about the relative contributions of the two terms to the equation for that type of system. For example, let us consider the specific case of the ratio of the acceleration of the wind (often called the *inertial term*) and the Coriolis term. We can write this ratio as follows:

$$\left|\frac{\partial u/\partial t}{(2\Omega v \sin\phi - 2\Omega w \cos\phi)}\right| \sim \frac{L/T^2}{2\Omega L/T}$$

where the relationship $\sim$ may be interpreted as 'scales as', and L, T, and Ω denote the orders of magnitude of the scales of motion. These are to be distinguished from the *dimensions*, which are written $[L], [T]$, and $[\Omega]$. Note that the trigonometric quantities have no dimension and an order of magnitude of 1. It is often the case that it is easier to measure the wind speed than to measure a time scale associated with the wind speed. Hence, it is more common to express such ratios in terms of the scales U and L, rather than L and T. So, we note that T has dimension $[L/U]$ and write the ratio as

$$\left|\frac{\partial u/\partial t}{(2\Omega v \sin\phi - 2\Omega w \cos\phi)}\right| \sim \frac{L/T^2}{2\Omega L/T}$$
$$\sim \frac{1}{2\Omega T}$$
$$Ro \sim \frac{U}{2\Omega L} \quad (5.4)$$

This ratio is known as the *Rossby number Ro*, named after Carl-Gustav Rossby (1898–1957), a scientist who made many important and fundamental contributions to the study of the atmosphere and ocean dynamics. When Ro is evaluated for a particular type of system it can tell us the importance of the Coriolis force in the development of the wind field.

Example Consider our case study storm at 12 UTC on Friday, 14 February 2003 (see Figure 4.2). In Limon, the wind report is 5 kts or 2.57 m s^{-1}, from the north. We estimate that the size of the system is around 1000 km across, and we know that $\Omega = 7.292 \times 10^{-5} \text{s}^{-1}$. In order to determine the Rossby number for this situation, we only need orders of magnitude for each of the dimensions in the expression, and we can see from a comparison to Table 5.1 that this system is similar to the typical scales for mid-latitude weather systems, as we might expect. Hence we can write

$$L \sim 10^6 \text{ m}$$

$$T \sim \text{several days} \sim 10^5 \text{ s}$$

$$\Rightarrow U \sim 10 \text{ m s}^{-1}$$

Such a system is called a *synoptic* scale system. We can also write

$$2\Omega \sim 10^{-4}$$

The Rossby number is then

$$Ro \sim \frac{U}{2\Omega L}$$

$$\sim \frac{10}{10^{-4} \times 10^6}$$

$$Ro \sim 0.1$$

Since the Rossby number is small (less than 1), this tells us that the Coriolis force, arising from the rotation of the frame of reference, is large compared to the acceleration of the wind, and hence must be important in the development of the flow around this system at this time. This is consistent with the result in the example in Section 4.4.2, which found that the Coriolis force was of the same order of magnitude as the pressure gradient force for this situation. However, our calculation here was much simpler to perform.

Let us consider some other illustrations of Rossby numbers. For example, the large, long-lived planetary waves that encircle the globe (Table 5.1) yield the following:

$$L \sim 10^7 \text{ m}$$

$$T \sim \text{weeks to months} \sim 10^6 \text{ s}$$

$$\Rightarrow U \sim 10 \text{ m s}^{-1}$$

$$Ro \sim 10^{-2}$$

This means that, even though the wind speed in a planetary wave is comparable to that in a mid-latitude weather system, the Coriolis force is comparatively even more important in the force balance that governs the planetary wave.

A foehn, a strong mountain lee wind, has very different scales:

$$L \sim 10^4 \, \text{m}$$
$$T \sim \text{hour} \sim 10^3 \, \text{s}$$
$$\Rightarrow U \sim 10 \, \text{m s}^{-1}$$
$$Ro \sim 10$$

In this case, the wind speed scale is again $10 \, \text{m s}^{-1}$, but because of the smaller spatial scales, we see that the Coriolis force is not particularly important in the generation of such winds.

Finally, let us consider the role of the Coriolis force in the flow of water out of a bath:

$$L \sim 1 \, \text{m}$$
$$T \sim 10 \, \text{s}$$
$$\Rightarrow U \sim 10^{-1} \, \text{m s}^{-1}$$
$$Ro \sim 10^3$$

So we see that the Coriolis force has an extremely small influence on the flow of water as it leaves the bath. This means that the hemisphere in which the bath is situated, which determines the direction of deflection induced by the Coriolis force, will have a negligible influence on the sense of rotation observed in the water flow. This is in contrast to mid-latitude weather systems, which have distinctive flows depending on the hemisphere – low-pressure systems rotate in the clockwise sense in the Southern Hemisphere and the counterclockwise sense in the Northern Hemisphere.

Other non-dimensional numbers that we may use in the course of our study include

$$\text{Reynolds number: } Re \sim \frac{\text{inertial}}{\text{viscous}} \sim \left| \frac{\partial u / \partial t}{\nu (\partial^2 u / \partial x^2 + \partial^2 u / \partial y^2 + \partial^2 u / \partial z^2)} \right|$$
$$\sim \frac{U^2/L}{\nu U/L^2} \sim \frac{UL}{\nu}$$
$$\text{Froude number: } Fr \sim \frac{\text{inertial}}{\text{gravity}} \sim \left| \frac{\partial u / \partial t}{g} \right| \sim \frac{U^2/L}{g} \sim \frac{U^2}{gL}$$

Thus, non-dimensional parameters like the Rossby, Reynolds, and Froude numbers allow an efficient measure of the relative importance of various terms in the dynamical equations, and aid the process of approximation and simplification. These parameters are also linked to the concept of *dynamic similarity*, in which flows at different spatial scales or with different fluid properties can be determined to be dynamically similar if *all* of their relevant non-dimensional parameters have the same magnitude. In such a situation, conclusions about one type of flow can be extended to other dynamically similar flows. Such a concept is indispensable when designing laboratory models that are used for understanding the behavior of atmospheric phenomena such as tornadoes.

5.4 Scale analysis

The process of simplifying the equations of motion using typical scales for the phenomenon of interest is called *scale analysis*. As an illustration of the technique, we will perform a scale analysis of the Navier–Stokes equations for a typical mid-latitude system. However, first we will make one prior simplification of the equation by examining the properties of the Coriolis force term.

5.4.1 Coriolis parameter

Recall the form of the Coriolis force (Equation (4.17)):

$$\frac{F_{Coriolis}}{m} = (2\Omega v \sin\phi - 2\Omega w \cos\phi)\,\vec{i} - 2\Omega u \sin\phi\,\vec{j} + 2\Omega u \cos\phi\,\vec{k}$$

We will define the Coriolis parameter, f, which arises from the fact that the important effects of the Earth's rotation in the middle latitudes arise mainly from the local vertical component of the rotation vector $\vec{\Omega}\bullet\vec{k} = \Omega\sin\phi$. We will see this is true once we perform a scale analysis (Section 5.4.2). Hence, we define

$$f = 2\Omega \sin\phi \tag{5.5}$$

For scaling purposes, we use the value of f at 45° latitude:

$$f_0 = 2\Omega \sin 45° = 1.03 \times 10^{-4} \tag{5.6}$$

Thus, the Rossby number (Equation (5.4)) is typically written

$$Ro \sim \frac{U}{f_0 L} \tag{5.7}$$

which does not change the analyses we performed in Section 5.3.

5.4.2 Scale analysis for a mid-latitude weather system

The Navier–Stokes equations are highly *nonlinear*, and cannot be solved analytically without considerable simplification. A nonlinear differential equation is one that includes products of terms involving the dependent variable – in this case, the

SCALE ANALYSIS

nonlinear terms are the advection terms, $u\partial u/\partial x$ and $v\partial u/\partial y$ for example. In nonlinear systems, different solutions cannot be superposed (added together) to form new solutions. This makes solving the equations much more difficult than linear systems. Hence, our approach in this book will be to simplify the equations in order to solve them.

Since much of the significant weather in middle latitudes is associated with cyclones and, to a lesser extent, anticyclones, we will derive a simpler form of the equation specifically for these systems. Typical values for the appropriate scales are shown in Table 5.2.

Then, the process of scale analysis is conducted as follows:

x-eqn	$\dfrac{\partial u}{\partial t}$	$+u\dfrac{\partial u}{\partial x}$	$+v\dfrac{\partial u}{\partial y}$	$+w\dfrac{\partial u}{\partial z}$	$=-\dfrac{1}{\rho}\dfrac{\partial p_d}{\partial x}$	$+\nu\left(\dfrac{\partial^2 u}{\partial x^2}+\dfrac{\partial^2 u}{\partial y^2}\right)$	$+\nu\dfrac{\partial^2 u}{\partial z^2}$	$+2\Omega v\sin\phi$	$-2\Omega w\cos\phi$
y-eqn	$\dfrac{\partial v}{\partial t}$	$+u\dfrac{\partial v}{\partial x}$	$+v\dfrac{\partial v}{\partial y}$	$+w\dfrac{\partial v}{\partial z}$	$=-\dfrac{1}{\rho}\dfrac{\partial p_d}{\partial y}$	$+\nu\left(\dfrac{\partial^2 v}{\partial x^2}+\dfrac{\partial^2 v}{\partial y^2}\right)$	$+\nu\dfrac{\partial^2 v}{\partial z^2}$	$-2\Omega u\sin\phi$	
scale	$\dfrac{U^2}{L}$	$\dfrac{U^2}{L}$		$\dfrac{UW}{H}$	$\dfrac{\delta p}{\rho L}$	$\dfrac{\nu U}{L^2}$	$\dfrac{\nu U}{H^2}$	$f_0 U$	$\approx f_0 W$
magnitude	10^{-4}	10^{-4}		10^{-5}	10^{-3}	10^{-16}	10^{-12}	10^{-3}	10^{-6}

Based on these magnitudes, we can see that the viscous term makes a very small contribution to the force balance experienced in a typical mid-latitude cyclone. So we can ignore that term with very little impact on the precision of our results. In fact, to a good degree of accuracy, we can also discard the Coriolis deflection caused by the vertical motion in the system $2\Omega w\cos\phi$, and the vertical advection term $w\partial u/\partial z$. Hence, we can rewrite the horizontal component of the Navier–Stokes equations in an approximate form:

$$\frac{\partial u}{\partial t}+u\frac{\partial u}{\partial x}+v\frac{\partial u}{\partial y}=-\frac{1}{\rho}\frac{\partial p_d}{\partial x}+2\Omega v\sin\phi$$
$$\frac{\partial v}{\partial t}+u\frac{\partial v}{\partial x}+v\frac{\partial v}{\partial y}=-\frac{1}{\rho}\frac{\partial p_d}{\partial y}-2\Omega u\sin\phi$$
(5.8)

Table 5.2 Typical scales for a mid-latitude weather system

Scale	Symbol	Magnitude
Horizontal wind scale	U	$10\,\text{m s}^{-1}$
Vertical wind scale	W	$10^{-2}\,\text{m s}^{-1}$
Horizontal length scale	L	$10^6\,\text{m}$
Vertical length scale (depth of troposphere)	H	$10^4\,\text{m}$
Time scale (L/U)	T	$10^5\,\text{s}$
Kinematic viscosity	ν	$10^{-5}\,\text{m}^2\,\text{s}^{-1}$
Dynamic pressure scale	$\delta p/\rho$	$10^3\,\text{m}^2\,\text{s}^{-2}$
Total pressure scale	P/ρ	$10^5\,\text{m}^2\,\text{s}^{-2}$
Gravity	g	$10\,\text{m s}^{-2}$
Density variation scale	$\delta\rho/\rho$	10^{-2}

We can write this in a shorter vector form if we define the horizontal wind to be $\vec{u}_h = u\vec{i} + v\vec{j}$ and the horizontal material derivative to be

$$\frac{D_h}{Dt} = \frac{\partial}{\partial t} + u\frac{\partial}{\partial x} + v\frac{\partial}{\partial y}$$

Then, we can write

$$\frac{D_h u}{Dt} = -\frac{1}{\rho}\frac{\partial p_d}{\partial x} + 2\Omega v \sin\phi$$

$$\frac{D_h v}{Dt} = -\frac{1}{\rho}\frac{\partial p_d}{\partial y} - 2\Omega u \sin\phi$$

$$\Rightarrow \frac{D_h \vec{u}_h}{Dt} = -\frac{1}{\rho}\left(\frac{\partial p_d}{\partial x}\vec{i} + \frac{\partial p_d}{\partial y}\vec{j}\right) - 2\vec{\Omega} \times \vec{u}_h$$

or

$$\frac{D_h \vec{u}_h}{Dt} = -\frac{1}{\rho}\left(\frac{\partial p_d}{\partial x}\vec{i} + \frac{\partial p_d}{\partial y}\vec{j}\right) - f\vec{k} \times \vec{u}_h \tag{5.9}$$

Turning to the vertical momentum equation, and using the total pressure form:

z-eqn	$\frac{\partial w}{\partial t}$	$+u\frac{\partial w}{\partial x} + v\frac{\partial w}{\partial y}$	$+w\frac{\partial w}{\partial z}$	$= -\frac{1}{\rho}\frac{\partial p}{\partial z}$	$+\nu\left(\frac{\partial^2 w}{\partial x^2} + \frac{\partial^2 w}{\partial y^2}\right)$	$+\nu\frac{\partial^2 w}{\partial z^2}$	$-g$	$+2\Omega u \cos\phi$
scale	$\frac{UW}{L}$	$\frac{UW}{L}$	$\frac{W^2}{H}$	$\frac{P}{\rho H}$	$\frac{\nu W}{L^2}$	$\frac{\nu W}{H^2}$	g	$f_0 U$
magnitude	10^{-7}	10^{-7}	10^{-8}	10	10^{-19}	10^{-15}	10	10^{-3}

Thus, the atmosphere in motion is strongly hydrostatic on the synoptic scale, since to an excellent degree of accuracy, we can write

$$\frac{1}{\rho}\frac{\partial p}{\partial z} = -g$$

Are the disturbances themselves hydrostatic? We find that if we do the same scale analysis using the dynamic pressure and the buoyancy term, there is a similar balance:

z-eqn	$\frac{\partial w}{\partial t}$	$+u\frac{\partial w}{\partial x} + v\frac{\partial w}{\partial y}$	$+w\frac{\partial w}{\partial z}$	$= -\frac{1}{\rho}\frac{\partial p_d}{\partial z}$	$+\nu\left(\frac{\partial^2 w}{\partial x^2} + \frac{\partial^2 w}{\partial y^2}\right)$	$+\nu\frac{\partial^2 w}{\partial z^2}$	$-\frac{(\rho - \rho_0)}{\rho}g$	$+2\Omega u \cos\phi$
scale	$\frac{UW}{L}$	$\frac{UW}{L}$	$\frac{W^2}{H}$	$\frac{\delta p}{\rho H}$	$\frac{\nu W}{L^2}$	$\frac{\nu W}{H^2}$	$\frac{\delta \rho g}{\rho}$	$f_0 U$
magnitude	10^{-7}	10^{-7}	10^{-8}	10^{-1}	10^{-19}	10^{-15}	10^{-1}	10^{-3}

So the answer is yes: synoptic scale disturbances themselves are strongly hydrostatic. This means that vertical velocities in such systems tend to be relatively small and are excited by departures from hydrostatic balance. One must always be careful to distinguish between the hydrostatic *equation* for an atmosphere at rest (Section 4.3.4) and the hydrostatic *approximation* for synoptic scale atmospheric motions derived here.

An important attribute of scale analysis is that in the process of deriving approximate equations for the system of interest, the level of accuracy is immediately quantified. Hence, one can choose different levels of approximation depending on the application. This choice may depend, for example, on the accuracy of the observations to be used, or on the need to retain certain terms in the equation.

5.5 The geostrophic approximation

In Section 5.3, we saw that the ratio of the inertial term and the Coriolis term in the Navier–Stokes equations is characterized by the Rossby number. We will now look at flows in which this ratio is very small. All flows in which $Ro \rightarrow 0$ are termed *geostrophic*; however, the definition of 'smallness' may vary with application. In our case we will assume that a 10% error is acceptable, and hence the case of a mid-latitude cyclone satisfies this criterion:

$$Ro = U/f_0 L = 10^{-1}$$

Then the horizontal momentum Equation (5.8) reduces to

$$-\frac{1}{\rho}\frac{\partial p_d}{\partial x} + 2\Omega v \sin\phi = 0$$
$$-\frac{1}{\rho}\frac{\partial p_d}{\partial y} - 2\Omega u \sin\phi = 0$$
(5.10)

Since time does not appear in this set of equations, this is a diagnostic relationship which cannot predict the evolution of the velocity field. We call this set of equations the geostrophic approximation for mid-latitude weather systems. In fact, for *any* flow, it is possible to define a horizontal velocity field called the *geostrophic wind*, which is a component of the total horizontal velocity field that satisfies exactly the force balance between the pressure gradient force and the Coriolis force. Such a geostrophic wind can be written, using Equation (5.5), as

$$u_g = -\frac{1}{\rho f}\frac{\partial p_d}{\partial y}$$
$$v_g = \frac{1}{\rho f}\frac{\partial p_d}{\partial x}$$
(5.11)

We can see the effect of this force balance in the presence of a simple meridional pressure gradient in Figure 5.1. It is clear from this diagram that the geostrophic wind must increase in magnitude in the presence of an increased pressure gradient, but that the Coriolis force will then increase concomitantly and thus maintain the geostrophic force balance. If such an increase occurs in time, of course it is not the geostrophic wind that can accelerate, since the equation governing this wind allows no time dependence.

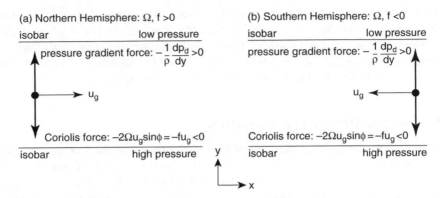

Figure 5.1 Geostrophic force balance and corresponding zonal wind in the case of a simple meridional pressure gradient for (a) the Northern and (b) the Southern Hemisphere

Example Recall the example explored in Sections 4.3.2 and 4.4.2, concerning the calculation of the pressure gradient force and Coriolis force in the vicinity of our case study low-pressure system. The map is reproduced in Figure 5.2 with the calculated forces illustrated by arrows. In this case, the pressure gradient force between Dodge City and Limon is directed from the high to the low pressure with a magnitude of $9.7 \times 10^{-5}\,\text{N}\,\text{kg}^{-1}$. The horizontal Coriolis force at Limon is directed to the right of the wind velocity vector at Limon with a magnitude of $2.4 \times 10^{-4}\,\text{N}\,\text{kg}^{-1}$. Hence, we can see that while there is a very approximate balance in the magnitudes of the forces, the directions of the forces do not suggest a strict geostrophic balance.

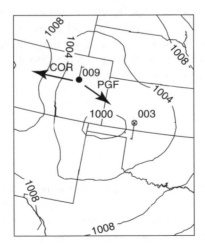

Figure 5.2 Sea level pressure over Oklahoma and surrounding states on 14 February 2003 at 12 UTC, showing the pressure and wind reports from Denver and Dodge City, and the geostrophic force balance

In part, this is due to the fact that we are using a wind field very close to the surface, where it is not accurate to assume that viscous forces are small (Chapter 10 discusses this in more detail). In the presence of viscous forces, surface winds are typically smaller in magnitude and changed in direction toward low pressure (typically around 20°) compared to the geostrophic wind. Further, if the system in question is propagating, as this one is to the east, there will be a component of wind in the direction of movement. Hence, in this real application it is clear that only a component of the wind field can be described as being in geostrophic balance.

Despite these shortcomings of the geostrophic approximation, the resulting force balance does explain an important attribute of flow around cyclones in the real atmosphere: that the wind flows in a generally counterclockwise direction in the Northern Hemisphere and a clockwise direction in the Southern Hemisphere. This direction is called *cyclonic* so that the hemisphere need not be specified. Conversely, the wind flow tends to be clockwise around an anticyclone in the Northern Hemisphere and counterclockwise in the Southern Hemisphere, and such a flow is termed *anticyclonic*.

The geostrophic approximation also helps to explain why weather systems move (most of the time) from west to east. Consider two columns of air – one in a polar air mass and the other in a maritime tropical air mass (Section 1.3.1). Assuming the *same surface pressure*, the pressure at a given altitude in the cold polar air mass will be less than that in the warm tropical column, because the thickness of the warm column is greater than that of the cold column (Figure 5.3). This creates a net horizontal pressure gradient force toward the cold air mass at this altitude, even though there is no net pressure gradient force at the surface. Then, regardless of whether we are considering the Northern Hemisphere (where the cold air mass will be to the north) or the Southern Hemisphere (where it will be to the south), a geostrophic

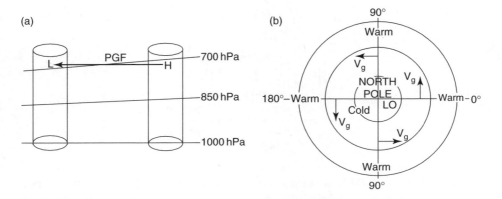

Figure 5.3 Geostrophic flow between a warm air mass and a cold air mass (a) leads to a westerly flow around the pole (b)

wind will be induced from the west. When we consider the entire polar region, rather than just one column, we see that the result is a large low-pressure area at some altitude above the surface, centered near the pole, with cyclonic westerly flow encircling it. This is the *polar vortex* and it is associated with a prevailing westerly flow. This upper level flow acts to steer disturbances, as we will see in upcoming chapters.

Example Calculate the thickness of the 1000 hPa to 700 hPa layer of the atmosphere in a polar air mass and in a maritime tropical air mass. Assume that the temperature in the polar air mass is $-20°C$ and that the temperature in the maritime tropical air mass is $20°C$.

We can use Equation (4.12) to calculate the thickness. As always, we must convert the temperature units to K.

For the polar air mass we find

$$\Delta Z = -\frac{R_d}{g_0^*} \int_{p_0}^{p} T d \ln p$$

$$= \frac{-287}{9.81} \int_{100\,000}^{70\,000} 253 d \ln p$$

$$= \frac{-287}{9.81} \times 253 \times \ln \frac{70\,000}{100\,000}$$

$$= 2640 \, \text{m}$$

And for the tropical air mass we find

$$\Delta Z = -\frac{R_d}{g_0^*} \int_{p_0}^{p} T d \ln p$$

$$= \frac{-287}{9.81} \int_{100\,000}^{70\,000} 293 d \ln p$$

$$= \frac{-287}{9.81} \times 293 \times \ln \frac{70\,000}{100\,000}$$

$$= 3057 \, \text{m}$$

Assuming that the 1000 hPa surface does not change elevation between the polar and tropical air masses, which is consistent with the surface pressure being the same in both locations, we find that the 700 hPa surface decreases in elevation by 417 m between the tropical and polar air masses. Assuming that these air masses are located at approximately 30°N and 60°N (or 3000 km apart) this gives a slope of $417 \text{ m}/3 \times 10^6 \text{ m} (= 1.39 \times 10^{-4})$.

Review questions

5.1 Verify that the Navier–Stokes equations (4.19) are dimensionally homogeneous.

5.2 Calculate the Rossby number for each type of motion listed in Table 5.1. You may assume a mid-latitude location for your calculations.

5.3 When cold air overlies a sloped ground surface (such as in the mountains) the cold air will flow down the slope under the influence of gravity. This *downslope* flow is known as a *katabatic* wind. Katabatic winds are observed over the world's great ice sheets in Greenland and Antarctica, as well as throughout smaller mountain ranges. Over the large ice sheets, the katabatic wind speed is of the order of 10 m s^{-1} and the horizontal scale of the flow is of the order of 1000 km. Over individual mountains katabatic wind speeds are of the order of 1 m s^{-1} and have a horizontal scale of 1 km.

(a) Calculate the Rossby number for both types of katabatic flow.

(b) Discuss the implications of the difference in Rossby number for the forces of relevance in each flow.

5.4 (a) What is the range of values for the Reynolds number for the types of motion listed in Table 5.1?

(b) What does this imply about the role of the viscous force for these types of motion?

(c) Under what conditions would the viscous force be important?

5.5 What physical conditions would result in a large Froude number and a small Froude number?

5.6 Assume that the scales given in Table 5.1 for mid-latitude weather systems are appropriate for tropical and polar weather systems as well.

(a) Calculate the Rossby number at latitudes 0°, 10°, 30°, 45°, and 90°.

(b) Based on your results for part (a), where can the geostrophic approximation be applied and with what accuracy?

5.7 For mid-latitude synoptic scale systems, what terms can be neglected in the horizontal momentum equation to give an accuracy of:

(a) 10%

(b) 1%

(c) 0.1%?

5.8 For mid-latitude synoptic scale systems, what terms can be neglected in the vertical momentum equation to give an accuracy of:

(a) 10%

(b) 1%

(c) 0.1%?

5.9 (a) Estimate the geostrophic wind speed at Little Rock, Arkansas at 00 UTC 15 Feb 2003 at the surface, 700 mb, 500 mb, and 300 mb levels using the station reports on the weather maps provided on the CD-ROM.

(b) How does the geostrophic wind speed compare to the observed winds at each of these levels?

5.10 Calculate the geostrophic wind at an altitude of 9.5 km and a latitude of 40°N for:

(a) Winter conditions with a surface temperature of 20°C at 30°N and −10°C at 50°N.

(b) Summer conditions with a surface temperature of 30°C at 30°N and 20°C at 50°N.

You may assume that the pressure at sea level is 1000 hPa and that the lapse rate is $6°C\,km^{-1}$.

6 Simple steady motion

6.1 Natural coordinate system

Typically, we use the latitude–longitude coordinate system that we introduced in Section 2.4 when describing atmospheric motions. However, for some classes of atmospheric flow it is more convenient to use another coordinate system. There are many choices of coordinate system on the Earth, including systems that use pressure instead of height as a vertical coordinate, and spherical coordinate systems that assume the Earth is a perfect sphere. In this chapter, we will make use of the *natural coordinate system*, which is a coordinate system in which one axis is everywhere tangent to the horizontal wind, represented by the unit vector $\vec{\tau}$, and a second axis is normal to and to the left of the wind, represented by unit vector $\vec{\eta}$ (Figure 6.1). We assume that the flow has a vertical component that is negligible or zero, and hence retain unit vector $\vec{k}$ as our vertical axis. This coordinate system has the potential to change in orientation with both time and space, but it follows that for all times and locations the wind field vector has a single component, directed in the $\vec{\tau}$ direction. We denote this wind field $\vec{V} = V\vec{\tau}$, and define (s, n, z) as our coordinate locations, which correspond to locations on the axes $\left(\vec{\tau}, \vec{\eta}, \vec{k}\right)$.

In order to write our Navier–Stokes equations in this coordinate system, we must first determine the form of the material derivative in this coordinate system, which means we must determine the form of

$$\frac{D\vec{V}}{Dt} = \vec{\tau}\frac{DV}{Dt} + V\frac{D\vec{\tau}}{Dt}$$

Based on the geometry shown in Figure 6.2, we can write

$$\delta\psi = \frac{\delta s}{R} \quad \text{and} \quad \delta\psi = \frac{|\delta\vec{\tau}|}{|\tau|} = |\delta\vec{\tau}|$$

$$\Rightarrow |\delta\vec{\tau}| = \frac{\delta s}{R}$$

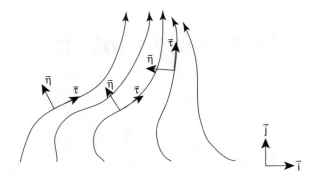

Figure 6.1 Natural coordinate system $(\vec{\tau}, \vec{\eta}, \vec{k})$ in a hypothetical wind field. The wind field everywhere is, by definition, $\vec{V} = V\vec{\tau}$

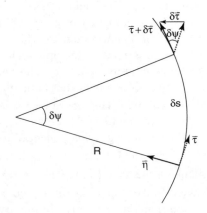

Figure 6.2 Change in orientation of the natural coordinate axes following a circular path with radius R. Adapted from Holton (1992)

Since, as $\delta s \to 0$, $\delta \vec{\tau}$ is parallel to $\vec{\eta}$,

$$\frac{d\vec{\tau}}{ds} = \frac{\vec{\eta}}{R}$$

$$\frac{D\vec{\tau}}{Dt} = \frac{D\vec{\tau}}{Ds}\frac{Ds}{Dt}$$

$$= \frac{\vec{\eta}}{R}V$$

$$\Rightarrow \frac{D\vec{V}}{Dt} = \vec{\tau}\frac{DV}{Dt} + \vec{\eta}\frac{V^2}{R}$$

Hence the acceleration of the parcel in this coordinate system is made up of two components, the rate of change of the parcel speed along the path of the flow, and the centripetal acceleration of the parcel if it follows a curved path. For this equation R will be positive when the center of curvature is in the positive $\vec{\eta}$ direction (to the left of the wind direction) and negative when the center of curvature is in the negative $\vec{\eta}$ direction (to the right of the wind direction).

Now we must consider the other terms in the Navier–Stokes equations. We know that the Coriolis force always acts normal to the direction of motion, and to the right in the Northern Hemisphere. Hence, it can be expressed simply as $-fV\vec{\eta}$, since f is positive in the Northern Hemisphere. In the Southern Hemisphere f is negative (due to the fact that ϕ is negative in the Southern Hemisphere). This results in a Coriolis term $(-fV\vec{\eta})$ that is positive in the Southern Hemisphere and directed to left of the wind as expected. The pressure gradient force, in general, will have components in the $\vec{\tau}$ and the $\vec{\eta}$ directions, and hence can be written as

$$-\frac{1}{\rho}\left(\frac{\partial p_d}{\partial s}\vec{\tau} + \frac{\partial p_d}{\partial n}\vec{\eta}\right)$$

We can also express the viscous force in the same way. However, for this chapter we will assume that we are considering motions that are far enough from the surface of the Earth that friction plays little role. The form of the Navier–Stokes equations that omits viscous forces is often called the Euler equation. In the natural coordinate system then, the horizontal momentum equations are

$$\text{s component:} \quad \frac{DV}{Dt} = -\frac{1}{\rho}\frac{\partial p_d}{\partial s}$$

$$\text{n component:} \quad \frac{V^2}{R} = -fV - \frac{1}{\rho}\frac{\partial p_d}{\partial n}$$

(6.1)

Since we have assumed a priori that there is no vertical motion, there is no vertical momentum equation and a simple hydrostatic balance pertains.

6.2 Balanced flow

A special case of purely horizontal frictionless flows, such as those governed by Equation (6.1), is known as *balanced flow*, which has the further restriction of being steady state – that is, all time derivatives are zero. It should be apparent that the geostrophic flow derived through scaling in the previous chapter is a type of balanced flow, since it is horizontal, does not change in time, is hydrostatic in the vertical, and experiences no viscous forces. With the assumption of steady state flow, Equation (6.1) reduces to

$$\text{s component:} \quad \frac{DV}{Dt} = -\frac{1}{\rho}\frac{\partial p_d}{\partial s} = 0$$

n component: $\quad\dfrac{V^2}{R} = -fV - \dfrac{1}{\rho}\dfrac{\partial p_d}{\partial n}$

This necessarily implies that all flows must be parallel to the isobars, since $\partial p_d/\partial s = 0$; that is, there is no change in pressure along the direction of flow. The single equation governing all balanced flows is then simply

$$\dfrac{V^2}{R} + fV = -\dfrac{1}{\rho}\dfrac{\partial p_d}{\partial n} \tag{6.2}$$

If we know something about the atmospheric system of interest, we can make this equation even simpler. We consider all the possibilities below.

6.2.1 Inertial oscillations

Consider a flow in which there is no significant pressure gradient present, or alternatively, there is a small pressure gradient that is balanced by viscous forces. In either case, such a flow can be treated as though there are no external forces acting on the air parcels, and they are simply under the influence of the rotating frame of reference. Thus, this kind of flow is as fundamental as straight motion at uniform velocity in a stationary frame of reference. Such flows, called inertial oscillations, are regularly observed in the lower atmosphere above the influence of the surface where frictional forces are significant.

The equation that governs inertial oscillations is then

$$\dfrac{V^2}{R} + fV = 0 \tag{6.3}$$

which allows us to solve for the speed of the air parcels as a function of the radius of curvature R and the latitude:

$$V = -fR$$

The period of the oscillation is then

$$T = \left|\dfrac{2\pi R}{V}\right| = \dfrac{2\pi}{f} = \dfrac{2\pi}{2\Omega \sin\phi} = \dfrac{1\ \text{day}}{2\sin\phi} \tag{6.4}$$

since the Earth undergoes one rotation (2π) per day. Hence, there is a characteristic period which can be identified in observations, an example of which is shown in Figure 6.3. The atmospheric motion was measured using a radar wind profiler, and is shown in the form of a *hodograph*, which plots u velocity on the x axis and v velocity on the y axis, as a function, in this case, of time. Such a graph would show a pure inertial oscillation as a perfect circle. The latitude of Whitewater is around 38°N, and hence the inertial period (time to complete one revolution) is around 19.5 hours. While a complete inertial circle was not observed, the oscillation in this observational period lasted around 17 hours.

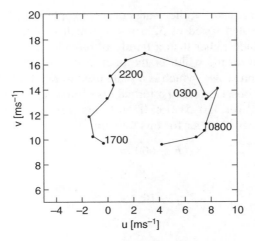

Figure 6.3 Hodograph showing an inertial oscillation of amplitude 4.19 m s^{-1} observed at Whitewater, Kansas (around 38°N) at an elevation of 192 m. The hours shown are local time. Reprinted from *Dynamics of Atmospheres and Oceans*, **33**, W. Blumen and J. K. Lundquist, 'Spin-up and Spin-down in rotating fluid exhibiting inertial oscillations and frontogenesis, 219–237, © 2001, with permission from Elsevier

6.2.2 Cyclostrophic flow

Consider now a flow where the horizontal scale is sufficiently small that we can neglect the Coriolis force. That is, we are considering flows in which the Rossby number is large. In fact, we can determine the Rossby number for such a flow by calculating the ratio of the first two terms in Equation (6.2):

$$Ro = \left|\frac{V^2}{R}\right| / |fV| = \frac{V}{|fR|}$$

A large cyclostrophic Rossby number implies a situation characterized by high wind speeds and tight rotation, such as a tornado or the even smaller phenomenon, a whirlwind (also known as a dust devil or willy willy). In this case, Equation (6.2) becomes

$$\frac{V^2}{R} = -\frac{1}{\rho}\frac{\partial p_d}{\partial n} \tag{6.5}$$

and hence we expect a force balance between the centrifugal force and the pressure gradient force. Theoretically, there is no limit to the strength of a system governed by this force balance, since no matter how strong the pressure gradient, the wind can increase to maintain the balance. In reality, however, frictional forces will tend to limit the intensity.

Example The most damaging tornado recorded in the United States was the so-called 'Tri-State Tornado' of 18 March 1925. It passed from Missouri to Illinois and thence to Indiana, causing the deaths of 695 people in just 4 hours. The tornado traveled at an estimated speed of 32.6 m s^{-1}, which is still considered a record. If it was a single tornado rather than a family of tornadoes, it would have been about 1200 m wide, based on the width of the path of destruction it left. It measured F5 on the Fujita–Pearson scale,[1] which is a scale used to rate tornado intensity based on the type of damage observed. An F5 tornado is estimated to have wind speeds in the range 117–142 m s^{-1} (or 420–512 km h^{-1}). Using this information, we can estimate a cyclostrophic Rossby number for this tornado:

$$R \sim 600 \text{ m}$$
$$V \sim 130 \text{ m s}^{-1}$$
$$f \sim 10^{-4} \text{ s}^{-1}$$
$$\Rightarrow Ro \sim 2.2 \times 10^3$$

This is certainly large enough to justify the neglect of the Coriolis term. The pressure gradient required to balance the estimated centrifugal force would be very large indeed:

$$\frac{\partial p_d}{\partial n} = -\rho \frac{V^2}{R}$$
$$\sim -1 \times \frac{130^2}{600}$$
$$\sim -28 \text{ Pa m}^{-1}$$

Given the difficulty of measuring the pressure gradient inside of a tornado such a large horizontal pressure gradient has never been observed in the atmosphere. Assuming that the pressure at the edge of the tornado was approximately 1000 hPa, this pressure gradient implies a central pressure in the tornado of 832 hPa! However, such a large pressure gradient is probably an overestimate, even for such an unusual event.

6.2.3 The geostrophic approximation

Now we consider the case where the Rossby number is very small. In that case, the ratio of the first two terms in Equation (6.2) suggests that we can neglect the centrifugal term in favor of the Coriolis term, and the governing equation reduces to

$$fV = -\frac{1}{\rho} \frac{\partial p_d}{\partial n} \tag{6.6}$$

This is simply geostrophic balance (Section 5.5) expressed in the natural coordinate system.

---

[1] Named after T. Theodore Fujita (1920–1998), a renowned tornado researcher at the University of Chicago, and Allen Pearson, director (now retired) of the National Severe Storms Forecast Center.

6.2.4 The gradient wind approximation

In situations where the Rossby number is close to 1, we must consider the three-way balance between Coriolis, centrifugal, and pressure gradient forces. Like geostrophic flow, it is always possible to define a component of the observed flow that satisfies Equation (6.2) – such a flow is called the *gradient wind*. We can determine the gradient wind in any situation by solving the quadratic equation for V:

$$\frac{V^2}{R} + fV + \frac{1}{\rho}\frac{\partial p_d}{\partial n} = 0$$

$$\Rightarrow V = -\frac{fR}{2} \pm \left(\frac{f^2 R^2}{4} - \frac{R}{\rho}\frac{\partial p_d}{\partial n}\right)^{\frac{1}{2}}$$

This can also be written in terms of the geostrophic wind (using Equation (6.6) and denoting the geostrophic wind as V_g):

$$V = -\frac{fR}{2} \pm \left(\frac{f^2 R^2}{4} + fRV_g\right)^{\frac{1}{2}} \qquad (6.7)$$

This equation clearly has several possible solutions depending on the signs of the radius of curvature and the pressure gradient, and the chosen sign of the root. However, we can use physical requirements to narrow the range of solutions. First, we can note that both the gradient wind and the geostrophic wind in the natural coordinate system must be positive definite. Tables 6.1 and 6.2 list the sign and magnitude for selected terms in the gradient wind equation for Northern and Southern Hemisphere flows respectively, for clockwise (CW) and counterclockwise (CCW) flow around areas of low (L) and high (H) pressure. From these tables we can find all of the possible physical solutions for this equation. Second, we can perform a Taylor series expansion on the expression given in Equation (6.7) – only those solutions in which $V \to V_g$ as the radius of curvature increases indefinitely have any physical meaning. We find that the solutions which conform to this requirement in the Northern Hemisphere are

cyclonic (CCW) flow around L: $R > 0$; $V = -\dfrac{fR}{2} + \left(\dfrac{f^2 R^2}{4} - \dfrac{R}{\rho}\dfrac{\partial p_d}{\partial n}\right)^{\frac{1}{2}}$

anticyclonic (CW) flow around H: $R < 0$; $V = -\dfrac{fR}{2} - \left(\dfrac{f^2 R^2}{4} - \dfrac{R}{\rho}\dfrac{\partial p_d}{\partial n}\right)^{\frac{1}{2}}$

and in the Southern Hemisphere are

cyclonic (CW) flow around L: $R < 0$; $V = -\dfrac{fR}{2} + \left(\dfrac{f^2 R^2}{4} - \dfrac{R}{\rho}\dfrac{\partial p_d}{\partial n}\right)^{\frac{1}{2}}$

anticyclonic (CCW) flow around H: $R > 0$; $V = -\dfrac{fR}{2} - \left(\dfrac{f^2 R^2}{4} - \dfrac{R}{\rho}\dfrac{\partial p_d}{\partial n}\right)^{\frac{1}{2}}$

Table 6.1 Sign and magnitude of terms in gradient wind equation for all possible flow regimes in the Northern Hemisphere

Term	Northern Hemisphere			
	Cyclonic (CCW) flow around L	Anticyclonic (CW) flow around H	Anticyclonic (CW) flow around L	Cyclonic (CCW) flow around H
f	+	+	+	+
R	+	−	−	+
$\dfrac{\partial p}{\partial n}$	−	−	+	+
$\left(\dfrac{f^2 R^2}{4} - \dfrac{R}{\rho}\dfrac{\partial p}{\partial n}\right)^{\frac{1}{2}}$	Always $> \dfrac{fR}{2}$	$< \dfrac{fR}{2}$ or imaginary for $\dfrac{f^2 R^2}{4} < \dfrac{R}{\rho}\dfrac{\partial p}{\partial n}$	Always $> \dfrac{fR}{2}$	$< \dfrac{fR}{2}$ or imaginary for $\dfrac{f^2 R^2}{4} < \dfrac{R}{\rho}\dfrac{\partial p}{\partial n}$
$-\dfrac{fR}{2}$	−	+	+	−
V positive for:	+ root only	Either root but $\dfrac{f^2 R^2}{4} > \dfrac{R}{\rho}\dfrac{\partial p}{\partial n}$	+ root only	never +

Table 6.2 Sign and magnitude of terms in gradient wind equation for all possible flow regimes in the Southern Hemisphere

Term	Southern Hemisphere			
	Anticyclonic (CCW) flow around L	Cyclonic (CW) flow around H	Cyclonic (CW) flow around L	Anticyclonic (CCW) flow around H
f	−	−	−	−
R	+	−	−	+
$\dfrac{\partial p}{\partial n}$	−	−	+	+
$\left(\dfrac{f^2 R^2}{4} - \dfrac{R}{\rho}\dfrac{\partial p}{\partial n}\right)^{\frac{1}{2}}$	Always $> \dfrac{fR}{2}$	$< \dfrac{fR}{2}$ or imaginary for $\dfrac{f^2 R^2}{4} < \dfrac{R}{\rho}\dfrac{\partial p}{\partial n}$	Always $> \dfrac{fR}{2}$	$< \dfrac{fR}{2}$ or imaginary for $\dfrac{f^2 R^2}{4} < \dfrac{R}{\rho}\dfrac{\partial p}{\partial n}$
$-\dfrac{fR}{2}$	+	−	−	+
V positive for:	+ root only	never +	+ root only	Either root but $\dfrac{f^2 R^2}{4} > \dfrac{R}{\rho}\dfrac{\partial p}{\partial n}$

BALANCED FLOW

The force balance represented by the Northern Hemisphere solutions is shown in Figure 6.4.

The cyclonic cases above always yield a real solution, but the anticyclonic cases require a restriction for real solutions to result:

$$\frac{f^2 R^2}{4} > \frac{R}{\rho} \frac{\partial p_d}{\partial n}$$

Hence, wind speeds near the center of anticyclones (high-pressure system) are generally light, whereas wind speeds associated with cyclones can be quite strong. This is because, in an anticyclone, the Coriolis force must balance the sum of the pressure gradient force and centrifugal force. The centrifugal force increases as V^2 but the Coriolis force increases only as V, which gives a physical reason for the upper limit in V. In a cyclone, the Coriolis and centrifugal forces act together to balance the pressure gradient, and hence the deeper the low, the faster the wind speed to create the balancing forces. In fact, the wind speed may become arbitrarily large in this model, although in reality surface frictional forces act to limit the intensity of the system, just as in the case of a tornado.

Example Consider our case study storm at 12 UTC on Friday 14 February 2003 (Figure 6.5). We can use the station observations to test the accuracy of the geostrophic and gradient wind models in this case. We start by estimating R, V, and the pressure gradient. The surface station report at Oklahoma City, Oklahoma gives a sea level pressure of 1004.4 hPa and a wind speed of 13 kts (from the actual weather observation at this time rather than the station model, which shows 15 kts on the surface weather map). This wind speed is then 6.7 m s^{-1} and is from the south-southwest. The sea level pressure reported near the center of the low is 1000.0 hPa, and at

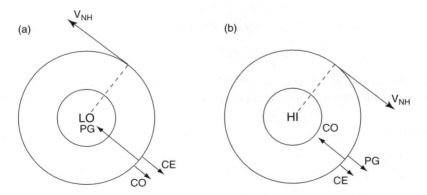

Figure 6.4 Force balance represented by the gradient wind (V_{NH}) model in the case of (a) a low-pressure system LO and (b) a high-pressure system HI in the Northern Hemisphere. The forces shown are the pressure gradient force (PG), the Coriolis force (CO), and the centrifugal force (CE)

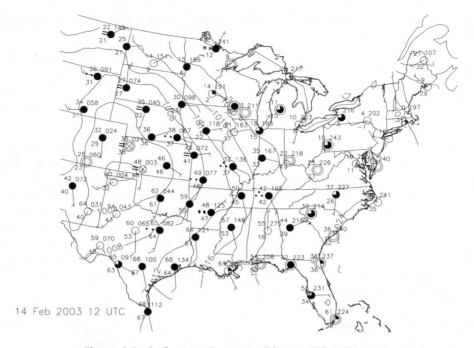

Figure 6.5 Surface weather map valid at 12 UTC 14 Feb 2003

this time the low is approximately 450 km to the northwest of Oklahoma City. From this, we can write

$$R = 4.5 \times 10^5 \text{ m}$$

$$\frac{\Delta p}{\Delta n} = \frac{(1000.0 - 1004.4) \times 10^2}{4.5 \times 10^5}$$

$$= -9.8 \times 10^{-4} \text{Pa m}^{-1}$$

$$V_{obs} = 6.7 \text{ m s}^{-1}$$

Oklahoma City is at a latitude of 35.2°N, which results in $f = 8.4 \times 10^{-5}$.

From this information, we can then calculate the geostrophic and gradient winds:

Geostrophic

$$V_g = -\frac{1}{f\rho}\frac{\partial p_d}{\partial n}$$

$$V_g = -\frac{1}{8.4 \times 10^{-5} \times 1.2} \times -9.8 \times 10^{-4}$$

$$= 9.7 \text{ m s}^{-1}$$

Gradient

$$V = -\frac{fR}{2} + \left(\frac{f^2 R^2}{4} - \frac{R}{\rho}\frac{\partial p_d}{\partial n}\right)^{\frac{1}{2}}$$

$$= -18.9 + \left(357 - 3.8 \times 10^5 \times -9.8 \times 10^{-4}\right)^{\frac{1}{2}}$$

$$= 8.1 \text{ m s}^{-1}$$

Hence, we can see that the gradient wind model gives a more reasonable result than the geostrophic wind model by taking account of the curvature of the isobars, which is significant so close to the center of the low. However, both models overestimate the observed wind speed due to the absence of frictional effects in the models. This serves to reduce the actual wind speed and turn the wind toward the center of the low. This is because friction reduces the wind speed sufficiently to reduce the Coriolis and centrifugal forces, so that they are no longer in balance with the pressure gradient force (Figure 6.6.) Hence, the actual wind will turn towards low pressure and may also accelerate. This, of course, represents a departure from balance.

It is possible to represent the relationship between the gradient wind and the geostrophic wind as a function of the Rossby number:

$$\frac{V^2}{R} + fV + \frac{1}{\rho}\frac{\partial p_d}{\partial n} = 0$$

$$\frac{V^2}{R} + fV - fV_g = 0$$

$$\Rightarrow V_g = V\left(1 + \frac{V}{fR}\right)$$

$$\therefore V_g = V(1 + Ro) \tag{6.8}$$

For mid-latitude synoptic systems, the difference between the gradient and geostrophic wind speeds generally does not exceed 10–20% (recall that the Rossby number for such systems is of order 10^{-1}). For tropical systems, in which the Rossby number is in the range 1–10, the gradient wind should be used in preference to the geostrophic approximation. In cyclones, the geostrophic wind overestimates the gradient wind ($R > 0$) and in anticyclones the geostrophic wind underestimates the gradient wind ($R < 0$).

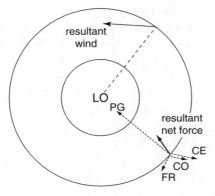

Figure 6.6 Force balance represented in the case of a low-pressure system in the presence of friction FR. All other forces indicated as in Figure 6.4

6.3 The Boussinesq approximation

One aspect of balanced flow that we have yet to consider is the appropriate form of the continuity equation. Let us consider specifically the geostrophic model for simplicity. The full continuity equation is

$$\frac{\partial \rho}{\partial t} + \left(u \frac{\partial \rho}{\partial x} + v \frac{\partial \rho}{\partial y} + w \frac{\partial \rho}{\partial z} \right) + \rho \left(\frac{\partial u}{\partial x} + \frac{\partial v}{\partial y} + \frac{\partial w}{\partial z} \right) = 0$$

If we perform a scale analysis on this equation using the scales in Table 5.2, we find that to an accuracy of the order of the Rossby number, Dines' compensation (see Section 3.7) holds:

$$\frac{\partial u}{\partial x} + \frac{\partial v}{\partial y} + \frac{\partial w}{\partial z} = 0$$

and it can be assumed that density is approximately constant in space and time. However, it can be shown that such a severe restriction results in a model for the wind field that does not allow any variation as we move up the atmospheric column – this is known as the *Taylor–Proudman theorem*. That is, the distribution of wind near the surface is also the distribution in the upper atmosphere. This is clearly not typically the case (see Figures 1.7 and 1.8, for example). Such a situation is known as having zero *wind shear*, which is expressed mathematically as

$$\frac{\partial u}{\partial z} = \frac{\partial v}{\partial z} = 0$$

Note that this condition says nothing about the vertical wind, only the variation of the horizontal wind in the vertical direction. Hence, we seek a form of the equations that allows the wind field to vary from the lower atmosphere to the upper atmosphere, but without allowing vertical motion or changes with time, which would complicate the picture unnecessarily. We can do this by allowing variations in density to give rise to buoyancy forces but have no impact on the horizontal force balance – this approach is known as the *Boussinesq approximation*, named after French physicist Valentin Boussinesq (1842–1929). It is constructed as follows.

First, we define a constant reference density ρ_{00} and a vertically varying density consistent with the hydrostatic pressure profile $\rho_0(z)$ as we did in Section 4.5.1. Then, we define a horizontally varying density $\rho(x, y)$ consistent with the horizontally varying dynamic pressure p_d. Note that we continue to allow no variations with time. Using this notation and recalling that we will not allow variations in density to impact the horizontal force balance, the geostrophic equation becomes

$$fV_g = -\frac{1}{\rho_{00}} \frac{\partial p_d}{\partial n}$$

Because, in applying the Boussinesq approximation, we wish to determine how the wind field changes in space, it is more appropriate to return to a coordinate system which is not tied to the wind field. So we write instead

$$fu_g = -\frac{1}{\rho_{00}}\frac{\partial p_d}{\partial y} \quad fv_g = \frac{1}{\rho_{00}}\frac{\partial p_d}{\partial x} \quad (6.9)$$

The vertical momentum equation in this approximation is of course the hydrostatic relationship which we will write in terms of the dynamic pressure and buoyancy force (see Section 5.4.2):

$$-\frac{(\rho-\rho_0)}{\rho_{00}}g - \frac{1}{\rho_{00}}\frac{\partial p_d}{\partial z} = 0 \quad (6.10)$$

where we have used the reference density so that variations in density only affect the calculation of the buoyancy force. The buoyancy force, denoted σ, can be expressed using the ideal gas equation in terms of temperature:

$$\sigma = -\frac{(\rho - \rho_0(z))}{\rho_{00}}g = \frac{(T - T_0(z))}{T_{00}}g \quad (6.11)$$

Combining Equations (6.10) and (6.11), we write

$$\sigma = \frac{1}{\rho_{00}}\frac{\partial p_d}{\partial z} \quad (6.12)$$

The Boussinesq approximation is an excellent one in the oceans where relative density differences nowhere exceed more than 1 or 2%. However, it is not strictly valid in the atmosphere. The reason is that air is compressible under its own weight to a degree that the density at the height of the *tropopause* (the boundary between the troposphere and the stratosphere) is only about one-quarter the density at sea level. At any particular height, however, departures of ρ from $\rho_0(z)$ are small and so we can often make use of the Boussinesq approximation for shallow motions.

In this case an appropriate form of the continuity equation is no longer strictly incompressible:

$$w\frac{\partial \rho_0}{\partial z} + \rho_0\left(\frac{\partial u}{\partial x} + \frac{\partial v}{\partial y} + \frac{\partial w}{\partial z}\right) = 0$$

Note that this retains the simplicity of a diagnostic rather than a prognostic system. Also, as we are about to discover, the flow is no longer strictly two dimensional.

6.4 The thermal wind

By taking the horizontal derivatives of Equation (6.12) and substituting these into the vertical derivatives of Equations (6.9), a new equation can be created which describes the variation of the geostrophic wind with height:

$$\frac{\partial u_g}{\partial z} = -\frac{1}{f}\frac{\partial \sigma}{\partial y} \quad \frac{\partial v_g}{\partial z} = \frac{1}{f}\frac{\partial \sigma}{\partial x}$$

or, by using Equation (6.11),

$$\frac{\partial u_g}{\partial z} = -\frac{g}{fT_{00}}\frac{\partial T}{\partial y} \qquad \frac{\partial v_g}{\partial z} = \frac{g}{fT_{00}}\frac{\partial T}{\partial x} \qquad (6.13)$$

This is known as the *thermal wind relationship*, and it connects the wind shear of the geostrophic wind (the vertical variation of the horizontal geostrophic wind) with the horizontal variation in temperature. It can also be understood as a balance between the buoyancy and the net Coriolis force. This relationship is illustrated as a movie on the accompanying CD-ROM.

Let us now consider a flow which, for the sake of illustration, is in the Northern Hemisphere and in a westerly direction, and where temperatures get progressively colder as we move north. The flow configuration is illustrated schematically in Figure 6.7 and for the real atmosphere in Figure 15.4. Thus, the relationship implied by Equation (6.13) correctly predicts the increase of zonal winds with height in the troposphere which is related to the decrease of temperature with latitude. The temperature gradient is reversed in the zone above the troposphere, known as the stratosphere: the polar stratosphere is actually warmer than the equatorial stratosphere, and the winds decrease with height in this layer, as predicted by the thermal wind relationship.

The thermal wind equation is, like the geostrophic equation, a diagnostic relation. As such it is useful in checking analyses of the observed winds and temperature fields for consistency. The thermal wind constraint is also important in ocean current systems wherever there are horizontal density contrasts.

6.4.1 Thermal advection

In the flow described in Figure 6.7, the geostrophic wind (and hence the isobars) are parallel to the isotherms. However, this is generally not the case. Consider the more typical situation in which the geostrophic wind blows at an angle to the horizontal isotherms. Suppose, for example, that the geostrophic wind at a particular height z blows from a region of high-temperature air to a region of low-temperature air – such

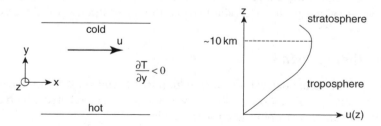

Figure 6.7 Illustration of the temperature–wind relationship embodied by the thermal wind equation

a situation is known as *warm advection* (Figure 6.8a). Then, the geostrophic wind at height $z + \Delta z$, where Δz is assumed to be small, can be written

$$\vec{u}(z + \Delta z) = \vec{u}(z) + \left(\frac{\partial u}{\partial z}\vec{i} + \frac{\partial v}{\partial z}\vec{j}\right)\Delta z + O\left(\Delta z^2\right)$$

$$\approx \vec{u}(z) - \frac{g}{fT_{00}}\frac{\partial T}{\partial y}\Delta z \vec{i}$$

because there is no temperature gradient in the x direction. Note that $\partial T/\partial y$ in this example is negative, and hence we add a vector in the positive x direction to determine the geostrophic wind at height $z + \Delta z$. Thus it follows, and this is true in general, that in the case of warm advection the geostrophic wind turns clockwise with height. We say that the wind *veers* with height. Conversely, the geostrophic wind turns counterclockwise, or *backs*, with height in the case of cold advection (Figure 6.8b). Note that the change in the wind vector is the same as in the warm advection case because the temperature gradient is the same: all that has changed is the direction of the geostrophic wind, which depends only on the orientation of the isobars.

In the Southern Hemisphere, these directions are of course reversed. A confusion that often occurs is that while the terms cyclonic and anticyclonic change sense with hemisphere (cyclonic is clockwise in the Southern Hemisphere and counterclockwise in the Northern Hemisphere), the terms veering and backing do not. Veering always means turning clockwise and backing anticlockwise, regardless of hemisphere. It is easier to remember that, regardless of hemisphere, warm advection leads to anticyclonic turning with height and cold advection leads to cyclonic turning with height of the geostrophic wind.

Example Figure 6.9 shows the air temperature distribution at 925 hPa for the case study storm at 00 UTC on 15 February 2003. This corresponds to the surface map shown in Figure 1.7. Also shown are the surface wind reports for two stations – Dodge City, Kansas and Nashville, Tennessee. At which station would one expect veering, and at which station would one expect backing?

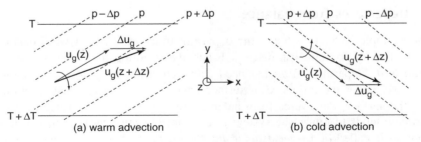

Figure 6.8 The temperature–wind relationship when the isotherms and isobars are not parallel

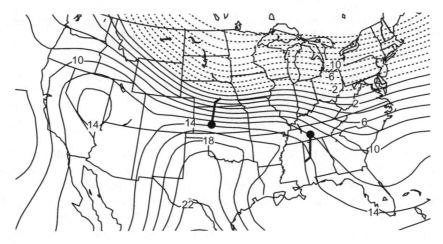

Figure 6.9 Contours of 925 hPa air temperature (in °C) valid at 00 UTC 15 Feb 2003. Also shown are the surface wind reports for Dodge City, Kansas and Nashville, Tennessee. The surface map valid at this time, which shows the full station reports for Dodge City and Nashville, is shown in Figure 1.7

The map shows a strong horizontal temperature gradient with temperature decreasing to the north. The isotherms are particularly close together in the vicinity of the fronts, as would be expected. Referring also to Figure 1.7, it is clear that at this time, Dodge City is located behind the advancing cold front, in the cold air mass. The wind is from the north at 5 kts, or 2.6 m s^{-1}. Since the wind is blowing toward warmer temperatures, this is a location of cold advection, and hence we would expect backing, or counterclockwise turning of the wind with height. Conversely, at this time Nashville is located ahead of the advancing warm front, and is experiencing a wind of the same magnitude but from the south. This results in warm advection and a veering of the horizontal wind with height. These conclusions can be verified using the upper air station reports from these locations, an exercise left for the student in the review questions.

6.5 Departures from balance

In the example in Section 6.2.4, the departure from the gradient wind balance due to the action of friction was discussed, and the resulting net force was illustrated (Figure 6.6). Departures from balanced flow can result in quite complex motions, particularly under the influence of surface friction, but for large-scale motions well above the surface, departures from balance are often very small since a process of *adjustment* tends to return the system to a balanced state. One example of this type of motion is called *quasi-geostrophic flow*, so named because it represents a small departure from the balanced case of geostrophic flow.

DEPARTURES FROM BALANCE

In quasi-geostrophic flow, the Rossby number is small, but finite, and both time evolution and vertical motion are allowed. We commence our analysis with the assumption that we are considering synoptic scale motions, and so we can use the synoptically scaled Equation (5.9):

$$\frac{D_h \vec{u}_h}{Dt} = -\frac{1}{\rho}\left(\frac{\partial p_d}{\partial x}\vec{i} + \frac{\partial p_d}{\partial y}\vec{j}\right) - f\vec{k} \times \vec{u}_h$$

as our starting point to derive a simplified horizontal momentum equation. With this scaling, we can also assume that the vertical momentum equation is simply the hydrostatic approximation – the appropriateness of this choice will be confirmed below.

The above equation is still rather complex, and we seek to simplify it based on our assumption that departures from geostrophy (and hence the Rossby number) are small. To do this, we can decompose the wind field into two components, the geostrophic flow and the departure from it, called the *ageostrophic wind*:

$$\vec{u}_h = \vec{u}_g + \vec{u}_a$$

and further, we define that

$$\frac{|\vec{u}_a|}{|\vec{u}_g|} \sim Ro \ll 1$$

We can then use the continuity equation to confirm the scaling for the allowed vertical motion. First we assume that variations in density are negligibly small; this has the implication, based on the derivation of the thermal wind equation, that any motion that begins from a state in which the geostrophic wind is independent of height will remain so. The appropriate continuity equation is

$$\frac{\partial u_h}{\partial x} + \frac{\partial v_h}{\partial y} + \frac{\partial w}{\partial z} = 0 \quad \text{and} \quad \frac{\partial u_g}{\partial x} + \frac{\partial v_g}{\partial y} = 0$$

$$\Rightarrow \frac{\partial u_a}{\partial x} + \frac{\partial v_a}{\partial y} + \frac{\partial w}{\partial z} = 0$$

$$\frac{\partial w}{\partial z} = -\left(\frac{\partial u_a}{\partial x} + \frac{\partial v_a}{\partial y}\right) \tag{6.14}$$

Thus, despite the fact that density variations are ignored, this model allows flow that is divergent in the horizontal. Consider the scales for the terms in this equation:

$$|u_g| \sim U \qquad |u_a| \sim RoU \qquad |w| \sim W$$
$$x, y \sim L \qquad z \sim H$$

Then Equation (6.14) implies that

$$\frac{W}{H} \lesssim \frac{RoU}{L}$$

We use $\leq$ because the right hand side of Equation (6.14) is a summation of terms that may be of different signs. Since for large-scale motions in the atmosphere $Ro \sim 0.1$ or less, and $H/L \sim 0.01$, it is clear that the vertical velocity scale W is many orders of magnitude less than the horizontal velocity scale U. Thus, small departures from geostrophy permit only very small vertical velocities, as we would expect.

Finally, we can use our scaling of the ageostrophic wind to derive a simplified form of the horizontal momentum equation. We will consider only the x component for simplicity:

$$\frac{\partial u}{\partial t} + u\frac{\partial u}{\partial x} + v\frac{\partial u}{\partial y} + w\frac{\partial u}{\partial z} = -\frac{1}{\rho}\frac{\partial p_d}{\partial x} + fv$$

$$\frac{\partial u}{\partial t} + u\frac{\partial u}{\partial x} + v\frac{\partial u}{\partial y} + w\frac{\partial u}{\partial z} = -fv_g + fv$$

$$\frac{\partial u_g}{\partial t} + \frac{\partial u_a}{\partial t} + (u_g + u_a)\frac{\partial (u_g + u_a)}{\partial x} + (v_g + v_a)\frac{\partial (u_g + u_a)}{\partial y} + w\frac{\partial (u_g + u_a)}{\partial z} = fv_a$$

Discarding terms of order Ro and smaller,

$$\frac{\partial u_g}{\partial t} + u_g\frac{\partial u_g}{\partial x} + v_g\frac{\partial u_g}{\partial y} = fv_a$$

We retain the Coriolis force arising from the ageostophic wind since

$$fv_a \sim fRoU \sim \frac{fU^2}{fL} \sim \frac{U^2}{L}$$

like the geostrophic advection terms. Thus, the governing equation for quasi-geostrophic flow is

$$\frac{D_g \vec{u}_g}{Dt} = -f\vec{k} \times \vec{u}_a \qquad (6.15)$$

6.5.1 Ageostrophic flow

As we have seen, the ageostrophic wind is the component of the wind that is present above and beyond the geostrophic wind, which flows parallel to the isobars. Equation (6.15) allows us to determine the magnitude of the cross-isobaric flow due to both changes in time and changes in space. Writing this equation as an expression to determine the ageostrophic flow,

$$u_a = -\frac{1}{f}\frac{\partial v_g}{\partial t} - \frac{1}{f}\left(u_g\frac{\partial v_g}{\partial x} + v_g\frac{\partial v_g}{\partial y}\right)$$

$$v_a = \frac{1}{f}\frac{\partial u_g}{\partial t} + \frac{1}{f}\left(u_g\frac{\partial u_g}{\partial x} + v_g\frac{\partial u_g}{\partial y}\right)$$

DEPARTURES FROM BALANCE

From this, we can see that in the Northern Hemisphere, the ageostrophic wind blows to the left of the acceleration vector $D_g \bar{u}_g/Dt$. Substituting Equation (6.9) into this gives the ageostrophic components in terms of the pressure gradient:

$$u_a = -\frac{1}{\rho_{00} f^2} \frac{\partial^2 p}{\partial x \partial t} - \frac{1}{\rho_{00} f^2} \left(u_g \frac{\partial^2 p}{\partial x^2} + v_g \frac{\partial^2 p}{\partial x \partial y} \right)$$

$$v_a = -\frac{1}{\rho_{00} f^2} \frac{\partial^2 p}{\partial y \partial t} - \frac{1}{\rho_{00} f^2} \left(u_g \frac{\partial^2 p}{\partial x \partial y} + v_g \frac{\partial^2 p}{\partial y^2} \right)$$

If the largest component of change in a particular flow is the change with time, then

$$u_a = -\frac{1}{\rho_{00} f^2} \frac{\partial^2 p}{\partial x \partial t}$$

$$v_a = -\frac{1}{\rho_{00} f^2} \frac{\partial^2 p}{\partial y \partial t}$$

(6.16)

which is known as the *isallobaric wind*, which flows normal to the isallobars, or lines of constant $\partial p_d/\partial t$. This concept was first introduced by Brunt and Douglas (1928), who suggested that isallobaric charts may be useful for determining regions of convergence and divergence.

In regions where the flow is approximately stationary in time, the cross-isobaric flow is represented by the advection term:

$$u_a = -\frac{1}{f} \left(u_g \frac{\partial v_g}{\partial x} + v_g \frac{\partial v_g}{\partial y} \right) = -\frac{1}{\rho_{00} f^2} \left(u_g \frac{\partial^2 p}{\partial x^2} + v_g \frac{\partial^2 p}{\partial x \partial y} \right)$$

$$v_a = \frac{1}{f} \left(u_g \frac{\partial u_g}{\partial x} + v_g \frac{\partial u_g}{\partial y} \right) = -\frac{1}{\rho_{00} f^2} \left(u_g \frac{\partial^2 p}{\partial x \partial y} + v_g \frac{\partial^2 p}{\partial y^2} \right)$$

(6.17)

and the ageostrophic wind is perpendicular to the advective acceleration.

Example Figure 6.10 illustrates these two components of the ageostrophic wind. In Figure 6.10(a) a map of isallobars (lines of constant pressure *tendency*) is shown. For this simple case we see that

$$\frac{\partial^2 p}{\partial x \partial t} = \frac{\partial}{\partial x} \frac{\partial p}{\partial t} < 0$$

and

$$\frac{\partial^2 p}{\partial y \partial t} = \frac{\partial}{\partial y} \frac{\partial p}{\partial t} = 0$$

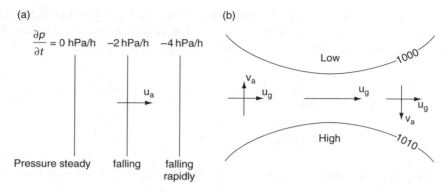

Figure 6.10 Zonal ageostrophic wind resulting from changes in pressure with (a) time and (b) space

Using this information in Equation (6.16), and assuming a Northern Hemisphere location ($f > 0$), we find that $u_a > 0$ and $v_a = 0$.

Figure 6.10(b) illustrates the ageostrophic component of the wind due to advective acceleration for a situation with isobars that are coming together (*confluence*) on the left side of the figure and isobars spreading apart (*diffluence*) on the right side of the figure. With this pressure pattern, the largest zonal geostrophic wind must be at the center of the figure, with weaker zonal geostrophic winds on the left and right sides, and $u_g > 0$ at all locations. It follows then that $\partial u_g/\partial x > 0$ on the left side of the figure and $\partial u_g/\partial x < 0$ on the right side of the figure. Applying this to Equation (6.17), again assuming a Northern Hemisphere location and neglecting meridional variations and v_g, we find that $u_a = 0$ and $v_a > 0$ on the left side of the diagram and $u_a = 0$ and $v_a < 0$ on the right side of the diagram. This cross-isobaric flow creates regions of convergence (upper left and lower right corners of Figure 6.10b) and divergence (lower left and upper right corners of Figure 6.10b) and thus vertical motion as diagnosed from Equation (6.14). We will see the importance of this small but crucial flow when we consider cyclogenesis in Chapter 9.

6.5.2 The maintenance of balance

As can be deduced from Figure 6.11, the Coriolis force associated with the cross-isobaric flow toward the surface low-pressure center is directed toward the west; that is, in the same direction as the geostrophic wind. In general, an ageostrophic wind directed toward low pressure will always tend to accelerate the wind in the direction of geostrophic flow. This is also true of the isallobaric component of the ageostrophic wind – it is the component of the flow that accelerates or decelerates to take up the geostrophic wind velocity consistent with the evolving pressure field. Hence, when a changing pressure field, either in space or in time, creates an imbalance of forces,

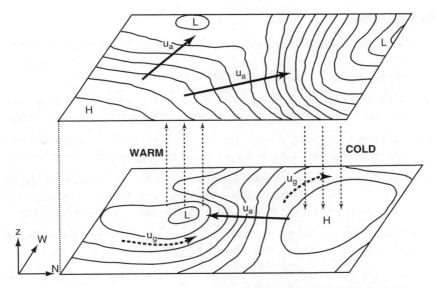

Figure 6.11 Vertical motion and subsequent circulation between the surface and the 500 hPa level induced by a thermal wind imbalance. Heavy dashed lines show the geostrophic wind and heavy solid lines show the ageostrophic wind. Vertical motion is indicated by fine dashed lines. Note that north is directed toward the right in this figure

a cross-isobaric flow is generated which will act to restore geostrophic balance. It is for this reason that, in the absence of other forcing, the departures from geostrophy in large-scale flow tend to be small. Such a process of restoring the flow to geostrophic balance is termed an *adjustment*.

A similar adjustment process takes place to maintain thermal wind balance. Consider flows occurring on two levels in the atmosphere, illustrated by our case study storm at 12 UTC on Monday, 17 February 2003, shown in Figure 6.11 as a schematic of the sea level pressure and 500 hPa height fields. Assume we are in a situation in which the flow is initially slower than required by geostrophic balance, and the wind shear is insufficient to maintain thermal wind balance with the meridional temperature gradient. The resulting cross-isobaric flow accelerates the geostrophic flow at both levels and increases the vertical wind shear. At the same time, the associated rising motion produces adiabatic cooling in the region of warm temperature and sinking motion produces adiabatic warming in the region of cold temperature, thus reducing the temperature gradient and hence the wind shear required to maintain thermal wind balance. In concert, these effects act to bring the flow back into balance with the pressure and temperature fields.

The rising motion in the warm air mass and the sinking motion in the cold air mass results in what is known as a *thermally direct* circulation. This process is highly efficient in the maintenance of thermal wind balance in large-scale motions in the atmosphere. In addition, the process itself acts to keep the ageostrophic flow small in comparison to the geostrophic flow.

Review questions

6.1 (a) What is the period of an inertial oscillation at the South Pole (latitude 90°S), Dunedin, New Zealand (latitude 46°S), Cape Town, South Africa (latitude 34°S), and Quito, Ecuador (latitude 0°)?

(b) What is the period of an inertial oscillation at latitude 46°N? How does this compare to the period of an inertial oscillation at Dunedin, New Zealand?

6.2 (a) Calculate the Rossby number for the atmospheric flows listed in Table 6.3. Assume a latitude of 40°N for all of the atmospheric flows considered.

(b) Calculate the change in pressure from the center to the edge of circulation for each of the atmospheric flows listed in Table 6.3 based on the equations for cyclostrophic flow, geostrophic balance, and gradient wind balance.

(c) Which balanced flow approximation is most appropriate for each type of flow and why?

(d) Describe physically why the change in pressure estimated in part (b) differs for each type of balanced flow.

6.3 What flow conditions (wind speed, radius of curvature, and Coriolis parameter) are required for the geostrophic approximation to provide a reasonable representation of the flow? (A qualitative answer is sufficient.)

6.4 At 15 UTC 28 Aug 2005 Hurricane Katrina was located at latitude 26.0°N and longitude 88.1°W in the Gulf of Mexico. Katrina was the third strongest hurricane on record in the Atlantic Basin, with a central pressure of 907 hPa. At this time reconnaissance aircraft measured a maximum wind speed at the surface of 75 m s^{-1} at a distance of 20 km from the center of the hurricane.

(a) Calculate the change in pressure between the center of the hurricane and the location of maximum wind speed assuming (i) cyclostrophic flow, (ii) geostrophic balance, and (iii) gradient wind balance.

(b) Which of the answers in part (a) is most likely to be closest to the actual change in pressure observed at this time? Why?

6.5 Draw a figure analogous to Figure 6.4 for a Southern Hemisphere location.

Table 6.3 Typical wind speed, radius, and direction of rotation for atmospheric flows in the Northern Hemisphere

Atmospheric flow	Wind speed (m s^{-1})	Radius (m)	Direction of rotation
Dust devil	10	10	CW or CCW
Tornado	100	500	CW or CCW
Hurricane	50	1×10^4	CCW
Synoptic cyclone	10	1×10^6	CCW

REVIEW QUESTIONS

6.6 Consider a balanced flow at 40°N where the magnitude of the pressure gradient force on a horizontally moving air parcel, expressed in natural coordinates, is 1.5×10^{-3} m s^{-2}.

(a) Find the geostrophic wind speed in m s^{-1}.

(b) Find the gradient wind velocity for a cyclone of this strength with radius of curvature 750 km.

(c) Find the maximum absolute value of the pressure gradient force for a balanced flow to exist in an anticyclone with a radius of curvature of 750 km. What is the corresponding geostrophic wind speed?

6.7 (a) Assume that the wind speeds given in Table 6.3 are equal to the gradient wind speed. Calculate the geostrophic wind speed for each type of flow using the gradient wind speed and the Rossby number.

(b) In which case does the geostrophic wind speed differ most from the gradient wind speed given in the table? In which case does it differ the least?

(c) Explain the physical reason for your answers to part (b).

6.8 Derive Equation (6.13) using Equations (6.9) and (6.12).

6.9 In a low-pressure system at 45°N, the geostrophic wind at the surface is 20 m s^{-1} from the south-west, and the geostrophic wind at 500 hPa is 25 m s^{-1} from the west. Is the wind veering or backing with height? What does this imply about thermal advection?

6.10 (a) Using the surface weather maps for the storm of 2003 on the CD-ROM, calculate the horizontal temperature gradient at the surface at Davenport, Iowa at 00 UTC 16 Feb 2003. (Hint: use the temperature observations from Chicago, Illinois (ORD), St. Louis, Missouri (STL), Des Moines, Iowa (DSM), and McCoy, Wisconsin (CMY) to calculate the temperature gradient.)

(b) What is the geostrophic wind at 850 hPa based on the temperature gradient calculated in part (a)? (Hint: you will need to calculate the geostrophic wind at the surface based on the sea level pressure observations at ORD, STL, DSM, and CMY.)

6.11 (a) Use the upper air weather maps for the storm of 2003 on the CD-ROM to estimate the geostrophic wind shear between 850 and 700 hPa at Oklahoma City, Oklahoma at 00 UTC 16 Feb 2003. You may assume that the winds reported on these weather maps are geostrophic winds.

(b) Calculate the horizontal temperature gradient required to produce the geostrophic wind shear calculated in part (a).

(c) Is the horizontal temperature gradient calculated in part (b) consistent with the temperatures shown on the 850 and 700 hPa upper air weather maps on the CD-ROM?

6.12 Use the weather observations at the surface and 850, 700, and 500 hPa on the CD-ROM for 00 UTC 15 Feb 2003 to verify that the wind backs with height at Dodge City, Kansas and veers with height at Nashville, Tennessee, consistent with the discussion for the example in Section 6.4.1.

6.13 A hodograph displays the change in wind speed and direction with height, giving a graphic depiction of vertical wind shear in a simple diagram.

(a) Using the data in Table 6.4 construct a hodograph using the blank hodograph diagram provided on the CD-ROM.

Remember that the wind direction given by meteorologists indicates the direction that the wind is coming from and that wind directions given by numerical values have the following geographic directions: $90° =$ east, $180° =$ south, $270° =$ west, and $360° =$ north, with intermediate numerical values corresponding to intermediate directions.

To make the hodograph, label the wind speed rings on the blank diagram with values appropriate for the data you are using. Then, for each height level, place a dot on the graph at the appropriate wind speed (given by the circles) and wind direction (given by the radial lines). Note that the wind directions plotted on the hodograph are the opposite to that which is normally used: a north wind is plotted at the bottom of the hodograph, and you proceed counterclockwise to east, south, and west. This convention is used so that the final plot on the hodograph indicates the direction the wind is going toward.

When you have placed all the dots, draw a line joining the dots from the surface point to the uppermost point.

(b) Assume the sounding was taken at a Southern Hemisphere location with latitude 44°S. Assume geostrophic and hydrostatic balance apply. Is the horizontal advection warm or cold in (i) the layer from the surface to 3 km

Table 6.4 Vertical profile of wind speed and direction observed at latitude 44°S

Altitude (km)	Wind speed (m s^{-1})	Wind direction (deg)
0.0	0.0	0
0.5	9.2	233
1.0	9.2	245
1.5	9.2	254
2.5	8.3	290
3.0	9.2	296
3.5	13	276
4.0	14	263
4.5	14	254
5.0	14	233

and (ii) the layer from 3 km to 5 km? Clearly explain the reasoning behind your answer.

(c) Calculate the amount of warm or cold air advection at 3 km in Kelvin per day. For simplicity, assume the average temperature is 273 K.

6.14 The mean temperature in a layer between 750 and 500 hPa decreases going toward the east by 3°C per 100 km. If the 750 hPa geostrophic wind is from the south-east at 20 m s^{-1}, what is the geostrophic wind speed and direction at 500 hPa? Let $f = 10^{-4}$ s^{-1}.

6.15 Rewrite Equation (6.15) as separate equations for both horizontal components of the wind, and expand the material derivative in this equation into local derivatives.

6.16 (a) What is the direction of the ageostrophic wind due to advective acceleration between Midland, Texas and Oklahoma City, Oklahoma at 300 hPa at 00 UTC 15 Feb 2003 based on the weather maps for the storm of 2003 provided on the CD-ROM? You may assume that the winds reported on this map are geostrophic winds.

(b) What is the direction of the ageostrophic wind due to advective acceleration between Oklahoma City, Oklahoma and Springfield, Missouri at this time?

6.17 What is the direction of the isallobaric component of the ageostrophic wind at the surface at Davenport, Iowa between 06 and 12 UTC 15 Feb 2003? Use the surface weather maps for 06 and 12 UTC provided for the storm of 2003 on the CD-ROM to answer this question.

7 Circulation and vorticity

7.1 Circulation

One way to describe an atmospheric flow that involves rotation or curvature is the *circulation*, which is defined as the line integral of the wind around a closed curve anywhere in the atmosphere (e.g. Figure 7.1):

$$C = \oint \vec{u} \bullet \vec{ds} \tag{7.1}$$

Unlike other measures of rotation in a flow, such as angular velocity, this definition does not require the center of rotation to be defined, and hence it is a convenient measure to use in situations where a single axis of rotation is difficult to identify. It is conventional to perform the integration in a counterclockwise direction around the curve and as a result C is positive for a counterclockwise flow.

7.1.1 Kelvin's circulation theorem

In general, the circulation around a particular closed curve will be a function of both time and space, since the velocity field is a function of time and space. So, if we wish to determine the rate of change of circulation, we can write

$$\frac{DC}{Dt} = \oint \frac{D\vec{u}}{Dt} \bullet \vec{ds}$$
$$= \oint -\frac{1}{\rho}\frac{\partial p}{\partial s}ds + \oint \frac{\partial \Phi}{\partial s}ds + \oint \text{ friction}$$

where we have neglected the rotation of the frame of reference (that is, $\vec{u}$ in this case is an absolute velocity relative to a fixed frame outside the Earth), and we have expressed the gravity force in the form of the *geopotential* (see Section 4.3.1). There is an additional term in this equation which arises from the advection of the closed curve itself, but it turns out to be zero anyway, so it has also been neglected.

Applied Atmospheric Dynamics   Amanda H. Lynch, John J. Cassano
© 2006 John Wiley & Sons, Ltd

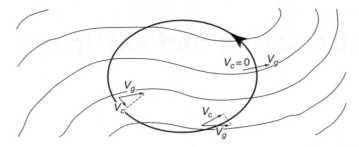

Figure 7.1 Components V_c of a gradient flow V_g that contribute to the circulation around the closed curve

Consider first the gravity term:

$$\oint \frac{\partial \Phi}{\partial s} ds = \oint d\Phi$$
$$= [\Phi_2 - \Phi_1]$$
$$= 0 \text{ around a closed circuit}$$

It is intuitive that the gravity force would not be involved in the generation of rotating motion, since gravity acts through the center of mass of any body.

Similarly, if the flow is *barotropic* (see Section 1.3.3), then density is a function only of pressure, and we can write

$$-\oint \frac{1}{\rho(p)} \frac{\partial p}{\partial s} ds = -\oint \frac{1}{\rho(p)} dp$$
$$= -\left[\frac{1}{\rho(p_2)} - \frac{1}{\rho(p_1)}\right]$$
$$= 0 \text{ around a closed circuit}$$

For the moment, let us also assume that the fluid is *inviscid*: that is, frictionless. From this, Kelvin's theorem that *circulation is constant in a barotropic, inviscid fluid* follows.

7.1.2 Bjerknes' circulation theorem

In the atmosphere, the conditions required for Kelvin's circulation theorem are rarely met, and the more important conclusion is that changes in circulation can arise from both viscous forces and baroclinicity (Section 1.3.3), as published by V. Bjerknes in 1937. We will consider the generation of circulation by friction in Chapter 10. The generation of a circulation by baroclinicity can be illustrated by considering the development of a sea breeze (illustrated in Figure 7.2 and shown as an animation

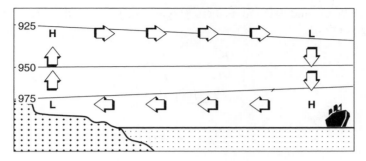

Figure 7.2 The configuration of isobars (shown in hPa) and circulation (arrows) when a sea breeze has reached a mature stage

on the accompanying CD-ROM). During a typical day at the beach, you may notice a steady wind blowing off the water in the late afternoon. This wind is generally noticeably cooler than the ambient air temperature. It is known as a sea breeze, and it occurs in response to differences in temperature between a body of water and neighboring land. That is, a sea breeze occurs in the presence of baroclinicity.

Overnight, the land and the ocean cool down from the previous day, and can approximate a common equilibrium temperature. If there is no significant weather passing through the region and the land surface near the coast is relatively flat, then there will be at most a very small pressure gradient between the land and the water. As the day progresses, the lower heat capacity of the land surface will cause it to increase in temperature more rapidly than the ocean in response to the same amount of solar energy impinging on the surface. Heat radiated back into the atmosphere from the land surface will cause warming of the overlying air, and expansion of the air column. This causes an increase in the thickness of the air column above the land relative to the air above the water, resulting in a horizontal pressure gradient aloft. This gradient can then initiate offshore flow at upper levels. Since the pressure at any location is determined by the weight of the air above it, the removal of air from higher levels causes the pressure at levels below to decrease. Hence, a pressure gradient in the reverse sense is created at lower levels, creating an onshore flow that we experience as a sea breeze. Rising and sinking air complete the circulation in response to continuity requirements.

Since a rising parcel of air cools and moistens, the inland arm of the circulation experiences conditions favorable for the formation of clouds. Hence, it is often possible to observe this branch of the sea breeze in the form of cumulus clouds, and sometimes even thunderstorms.

Example We can calculate an expected sea breeze based on a given temperature difference as follows. First, we apply the circulation theorem around a vertical circuit along the 975 mb and 925 mb isobars (Figure 7.2), using the ideal gas law and neglecting friction:

$$\frac{DC}{Dt} = -\oint \frac{1}{\rho} dp$$

$$= -\oint \frac{RT}{p} dp$$

$$= -\oint RT d\ln p$$

There is a contribution to the circulation only by the vertical segments because the horizontal segments are taken at constant pressure. The resulting rate of increase in the circulation is

$$\frac{DC}{Dt} = R\ln\left(\frac{p_b}{p_a}\right)(\overline{T}_2 - \overline{T}_1)$$

Taking $(\overline{T}_2 - \overline{T}_1)$ to be around 20°C, the rate of increase of circulation around this circuit is

$$\frac{DC}{Dt} = 287 \times \ln\left(\frac{92\,500}{97\,500}\right) \times 20$$

$$= -302 \text{ m}^2\text{s}^{-2}$$

We expect the result to be negative since the temperature difference is inducing a clockwise circulation.

The mean acceleration can be computed if we know the distance around the circuit and the mean temperature in the atmospheric column. We will assume that the mean temperature is about 15°C, and the horizontal distance 20 km. Sea breeze circulations can typically penetrate inland a maximum of around 40 km from shore. This is due to the increased surface friction, relative to the ocean surface, resulting from the topography of the land. To calculate the vertical distance, recall the equation for thickness (Equation (4.12))

$$\Delta Z = \frac{R\overline{T}}{g}\ln\left(\frac{p_a}{p_b}\right) = \frac{287 \times 288}{10} \times \ln\left(\frac{97\,500}{92\,500}\right) = 434\,\text{m}$$

Then

$$\frac{DC}{Dt} = \oint \frac{D\vec{u}}{Dt} \cdot d\vec{s}$$

$$\approx \frac{D\vec{u}}{Dt} \cdot \oint d\vec{s}$$

$$\Rightarrow \frac{Du}{Dt} \approx \frac{DC}{Dt}/(2 \times \Delta Z + 2L) \approx -7.39 \times 10^{-3}\,\text{m}\,\text{s}^{-2}$$

In the absence of retarding forces this would produce a wind speed of 25 m s^{-1} (just under 50 kts) in about an hour. In reality, as wind speed increases, the frictional force near the surface increases, retarding the acceleration. In addition, *temperature advection* (see Section 2.8) reduces the land–sea temperature contrast so that a balance is obtained.

7.1.3 Observing sea breezes

Sea breezes are most common during the sunny days of spring and early summer, since they require effective cooling at night followed by consistent heating during the day. The leading edge of a sea breeze is characterized by decreased temperature and rising motion, and hence has much in common with cold fronts (Section 9.1). The sea breeze front can be identified in observations of clouds and in satellite imagery by a line of fair weather cumulus clouds landward of clear air (see the CD-ROM for an example). The sinking motion and lower temperatures over the ocean prevent clouds from forming. Sea breezes can also occur in the vicinity of large lakes.

7.1.4 Relative circulation

So far, we have neglected the fact that we are in a rotating frame of reference. On the Earth, a component of the circulation around any circuit will be due to the rotation of the frame, that is

$$C_{absolute} = C_{earth} + C_{relative}$$

The circulation due to the rotation of the Earth is then simply

$$C_{earth} = \oint \vec{u}_{earth} \bullet d\vec{s}$$
$$\approx R\Omega \times 2\pi R \sin \phi$$
$$C_{earth} \approx 2\pi \Omega R^2 \sin \phi$$

where R is the radius of the Earth and Ω is the angular velocity of the Earth. The primary importance of this component is the fact that it is dependent upon latitude. Hence, a meridional flow of air under conditions of conserved absolute circulation will experience an induced relative circulation to compensate for the change in the circulation due to the rotation of the Earth. For example, consider the trade winds over the Atlantic Ocean in northern summer (Figure 7.3). Trade winds flowing north and south converge at the latitude of maximum heating which is around 10°N in this figure. The southerly trades, originating in the Southern Hemisphere, acquire a clockwise (that is, negative) rotation as they cross the equator and the circulation due to the Earth's rotation increases.

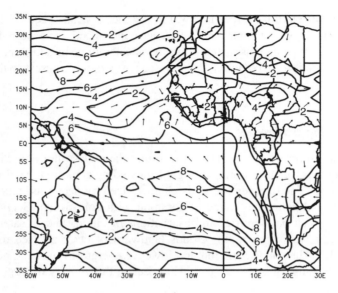

Figure 7.3 Average 1000 hPa vector wind (m s^{-1}) for August 2003 for the tropical Atlantic. The contours show the magnitude of the wind and the vectors show the wind direction. NCEP Reanalysis data provided by the NOAA-CIRES Climate Diagnostics Center, Boulder, Colorado, from its Web site at http://www.cdc.noaa.gov/

7.2 Vorticity

Another way to describe the curved motion of fluid parcels without reference to a center of rotation is a quantity known as the *vorticity*, which is simply the circulation per unit area:

$$\zeta = \frac{\partial C}{\partial A} \tag{7.2}$$

This definition simplifies the analysis since, if we can calculate ζ, we no longer need to define a specific closed circuit in the flow of interest. Since the circulation represents the *flux* of vorticity through a specified area closed circuit, the circulation is often termed the *vortex strength*.

The calculation of vorticity is particularly simple in the case of solid body rotation. Consider an infinitesimally small disk of radius δr rotating with angular velocity ω corresponding to a speed V at the rim:

$$\zeta = \frac{\delta C}{\delta A} = \frac{2\pi \delta r V}{\pi \delta r^2} = \frac{2V}{\delta r} = 2\omega$$

If we expand this disk to a finite area, we find that components of the circulation on opposite sides of the disk cancel out, and hence the relationship $\zeta = 2\omega$ is generally true for all solid body rotation. It follows that the vorticity of any fluid parcel on the

Earth has a component due to the solid body rotation of the Earth, at any point of which $\omega = \Omega \sin \phi$ and hence

$$\zeta_{earth} = 2\Omega \sin \phi = f$$

Therefore ζ_{earth} is positive in the Northern Hemisphere and negative in the Southern Hemisphere.

We can derive the relative vorticity of a fluid in motion by calculating the circulation around a small rectangular fluid element, as shown in Figure 7.4, where we have neglected motion in the vertical:

$$C = \oint \vec{u} \cdot d\vec{s} = \oint (u\,dx + v\,dy)$$

and hence the circulation around the infinitesimal fluid element ABCD:

$$\delta C = C_{AB} + C_{BC} + C_{CD} + C_{DA}$$
$$= u\delta x + \left(v + \frac{\partial v}{\partial x}\delta x\right)\delta y - \left(u + \frac{\partial u}{\partial y}\delta y\right)\delta x - v\delta y$$
$$= \frac{\partial v}{\partial x}\delta x \delta y - \frac{\partial u}{\partial y}\delta y \delta x$$
$$= \left(\frac{\partial v}{\partial x} - \frac{\partial u}{\partial y}\right)\delta A$$

where δA is the area of ABCD. Then the relative vorticity is simply

$$\zeta = \frac{\delta C}{\delta A}$$
$$\zeta = \frac{\partial v}{\partial x} - \frac{\partial u}{\partial y} \qquad (7.3)$$

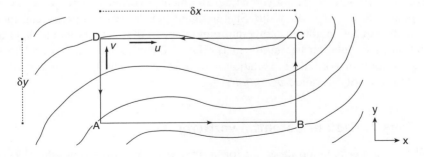

Figure 7.4 Calculation of circulation around a rectangular fluid element

The relative vorticity will be positive for a counterclockwise rotation and will be negative for a clockwise rotation. Therefore, for flow around a low-pressure center in the Northern Hemisphere, the relative vorticity will be positive. In the Southern Hemisphere the relative vorticity for flow around a low-pressure center is negative.

Example In August 1992, a category 5 hurricane, named Andrew, hit southern Florida and the Louisiana coast. Hurricane Andrew caused 61 deaths and was one of the costliest hurricanes in US history up to that date (a satellite image is available on the CD-ROM). At the time the satellite image was taken, wind gusts up to 150 kts were observed at a distance from the center of 20 km. The vortex strength at this time was then

$$C = \oint \vec{u} \cdot d\vec{s}$$
$$U = \vec{u} \cdot d\vec{s} = 150 \text{ kts} = 77 \text{ m s}^{-1}$$
$$\Rightarrow C = 2\pi R U = 2\pi \times 20 \times 10^3 \times 77$$
$$\therefore C = 9.7 \times 10^6 \text{ m}^2 \text{ s}^{-1}$$

The vorticity within the hurricane is roughly

$$\zeta \approx \frac{C}{A} = \frac{9.7 \times 10^6}{\pi R^2}$$
$$\zeta \approx 7.7 \times 10^{-3} \text{ s}^{-1}$$

Note that this value is two orders of magnitude larger than the background vorticity due to the rotation of the Earth, which at latitude 25°N is $2\Omega \sin \phi = 3.1 \times 10^{-5} \text{ s}^{-1}$. This is typically true of all moderate to strong circulations on the Earth, although it does not imply that the background rotation is not important, as we will see in the next section.

In fact, although the above calculation is straightforward and illuminating as far as relative magnitudes are concerned, the most important aspect of relative vorticity is not its value at a particular time, but its change in space or time. Temporal changes in relative vorticity tell us about cyclone development – vorticity increases as cyclones spin up, and decreases as they die. Spatial changes in relative vorticity can indicate the influence of mountains or temperature gradients – the factors that can lead to the generation of circulation according to Bjerknes' theorem. One way to address the spatial variation of vorticity in isolation is to develop a quantity that is conserved in time – that quantity is called *potential vorticity*.

7.3 Conservation of potential vorticity

Recall Kelvin's circulation theorem: for an inviscid, barotropic flow, the circulation is conserved. We also saw that in most real cases, circulation (or vortex strength) changes due to the presence of baroclinicity or friction. However, we can simplify this situation by making two assumptions: that we are far from the surface and hence frictional effects are minimal, and that motion is adiabatic (that is, the potential temperature is constant). Such a situation is often called *potentiotropic*, because on a constant potential temperature surface, the density is a function of pressure alone:

$$\rho = \frac{p}{RT} = \frac{p}{R\theta}\left(\frac{p_0}{p}\right)^{R_d/c_p} = p^{c_v/c_p}\left(\frac{p_0^{R_d/c_p}}{R\theta}\right)$$

This is analogous to a barotropic fluid, in which the density is a function only of pressure everywhere in the fluid. However, adiabatic flow is more closely approximated in the real atmosphere than barotropic flow. Thus, on a constant potential temperature surface, the pressure gradient term does vanish and the fluid satisfies Kelvin's circulation theorem. Mathematically, this is written

$$\left.\frac{DC_{absolute}}{Dt}\right|_\theta = 0 \qquad (7.4)$$

indicating that the circulation must be evaluated on a closed loop that lies entirely on the constant potential temperature surface.

Using Equation (7.2) with the implied assumption that the constant potential temperature surface is approximately horizontal, we can write

$$C_{absolute} = \iint_A (\zeta_\theta + f)\, dA$$

$$\zeta_\theta + f = \lim_{\delta A \to 0} \frac{C_{absolute}}{\delta A}$$

$$\delta A\,(\zeta_\theta + f) = \text{constant} \qquad (7.5)$$

Suppose that a parcel is confined between potential temperature surfaces θ and $\theta + \delta\theta$, which are separated by a pressure interval δp. Because the parcel is confined to a given potential temperature ($\pm\frac{1}{2}\delta\theta$), the motion is by definition adiabatic. The mass of the parcel, given by $\delta M = \rho\,\delta z\,\delta A = -(\delta p/g)\,\delta A$, must be conserved following the motion. Hence we can write

$$\delta A = -\frac{\delta M g}{\delta p} \equiv -\frac{\delta M g}{\delta p} \times \frac{\delta\theta}{\delta\theta}$$

$$\delta A = -\frac{\delta M g}{\delta\theta}\left(\frac{\delta\theta}{\delta p}\right)$$

Since δM and $\delta\theta$ are constant, this can be simplified to

$$\delta A = -\text{constant} \times g\frac{\delta\theta}{\delta p} \qquad (7.6)$$

where we have retained g by convention. Eliminating δA between Equations (7.5) and (7.6) yields

$$\left(\text{constant} \times -g\frac{\delta\theta}{\delta p}\right)(\zeta_\theta + f) = \text{constant}$$

$$\therefore P = -g\frac{\delta\theta}{\delta p}(\zeta_\theta + f) = \text{constant}$$

where we have defined a new quantity P, conserved in adiabatic frictionless flow, which is known as the Rossby–Ertel *potential vorticity*. The term 'potential vorticity' is used in connection with several mathematical expressions. The physical meaning, however, is that any quantity termed potential vorticity is always in some sense a measure of the ratio of the absolute vorticity to the effective depth of the vortex. The concept was introduced by C. G. Rossby in 1936 and shown in the form similar to the one above by H. Ertel in 1939. The potential vorticity is a powerful concept because it is an integration of small-scale fluid element properties and large-scale flow properties.

Example A typical weather pattern when the prevailing flow impinges on a long mountain chain that is normal to the flow is shown in Figure 7.5, which took place as our case study storm ravaged the east coast. The Rocky Mountains are arrayed

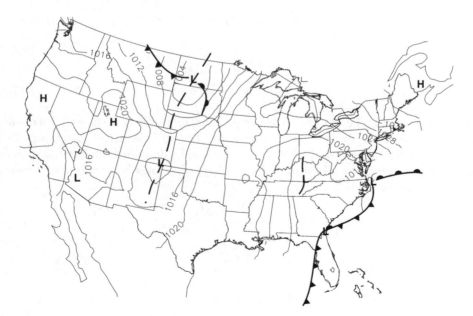

Figure 7.5 Mean sea level pressure map, showing low- and high-pressure centers, fronts, and troughs, at 00 UTC 17 Feb 2003

north–south through the western third of the United States, and the prevailing flow is westerly – recall from Chapter 6 that the thermal wind balance in the presence of the decreasing equator-to-pole temperature gradient requires that the balanced air flow be westerly. From the map, it is clear that high-pressure areas dominate to the west of the Rocky Mountains, and a trough of low pressure sits on the lee side, along with two closed low-pressure circulations. Why would such a configuration be typical?

Let us assume that we have conditions that approximate adiabatic, frictionless flow. Consider what happens when that flow impinges on a mountain barrier from the west (Table 7.1).

As the column of air approaches the mountain, the upper potential temperature surface rises, causing the column to increase in depth. This causes a decrease in $\delta\theta/\delta p$, and for conservation, this requires $(\zeta_\theta + f)$ to increase. The only way in

Table 7.1 Modifications to vortex depth and vorticity as a westerly flow impinges on an isolated mountain barrier in the Northern Hemisphere

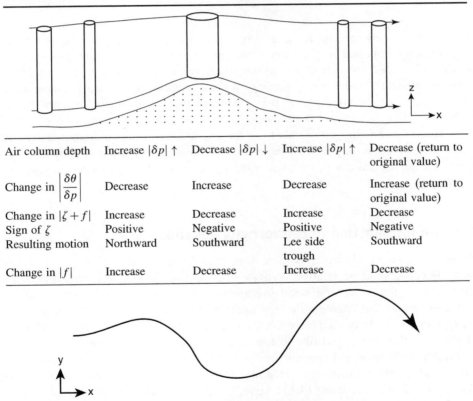

Air column depth	Increase $\|\delta p\| \uparrow$	Decrease $\|\delta p\| \downarrow$	Increase $\|\delta p\| \uparrow$	Decrease (return to original value)
Change in $\left\|\dfrac{\delta\theta}{\delta p}\right\|$	Decrease	Increase	Decrease	Increase (return to original value)
Change in $\|\zeta + f\|$	Increase	Decrease	Increase	Decrease
Sign of ζ	Positive	Negative	Positive	Negative
Resulting motion	Northward	Southward	Lee side trough	Southward
Change in $\|f\|$	Increase	Decrease	Increase	Decrease

which the flow can respond to effect such an increase is to introduce a rotation that causes ζ_θ to increase; that is, a cyclonic or counterclockwise rotation. This causes the flow to turn slightly northward, which, in turn, causes an increase in the planetary vorticity, f. By this time, the flow is impinging on the rising slopes of the mountain, causing the air column to decrease in depth. This causes a rapid increase in $\delta\theta/\delta p$ requiring a corresponding decrease in $(\zeta_\theta + f)$. This generates a strong anticyclonic rotation – the source of the typical high pressure on the windward side of a long mountain range. The resulting southward flow causes the planetary vorticity to decrease significantly.

After cresting the mountain, the vortex depth starts to increase rapidly again, causing the generation of cyclonic flow and the acquisition of increasing planetary vorticity. This is evident at the surface as a lee side trough, as we see in Figure 7.5. At this point, the air column is likely to have reached its original value, and if no further mountain ranges lie in the path of the air flow, one might expect no further changes in vorticity. However, the air column now lies to the north of its original position, and so with a larger value of f than initially, conservation requires $\zeta_\theta < 0$. This causes clockwise rotation and equatorward motion. This interplay between relative and planetary vorticity can continue downstream of the mountain, generating a lee side *wave*.

The interplay between the relative vorticity of the fluid elements and the planetary vorticity imparted to the large-scale flow results in circulation patterns such as that illustrated by Figure 7.5. As we will see in Section 8.2, this 'trade-off' between planetary and relative vorticity results in motions that are fundamental to the dynamics of the atmosphere. It is also clear that the depth of the flow has an influence – since variations in atmospheric depth are often associated with variations in temperature (Section 4.3.4), this links the important concepts of baroclinicity with vorticity generation. In general, it is useful to remember that the stretching of vortex tubes causes acquisition of cyclonic vorticity, and shrinking produces anticyclonic vorticity.

7.4 An introduction to the vorticity equation

From the example above, it is clear that even in situations where potential vorticity is conserved, the relative vorticity itself is not. In fact, it is this aspect that gives rise to the interesting weather patterns we see in the vicinity of mountains. We also saw in that example that the interplay between relative and planetary vorticity can be an important aspect of the generation of atmospheric circulations. Because of this, we would like to have an equation that we can use to predict the change in both time and space of the vorticity – this is called the *vorticity equation*. A complete form of this equation can be derived from first principles using the Navier–Stokes equations (4.19). Here, the vorticity equation for a synoptic scale mid-latitude weather system will be derived, based on the momentum equation for such a system (Equation (5.9)) and using the definition of vertical relative vorticity (Equation (7.3)):

$$\frac{D_h \vec{u}_h}{Dt} = -\frac{1}{\rho}\left(\frac{\partial p_d}{\partial x}\vec{i} + \frac{\partial p_d}{\partial y}\vec{j}\right) - f\vec{k} \times \vec{u}_h$$

$$\Rightarrow \frac{D_h u}{Dt} = -\frac{1}{\rho}\frac{\partial p_d}{\partial x} + fv \quad (1) \qquad \frac{D_h v}{Dt} = -\frac{1}{\rho}\frac{\partial p_d}{\partial y} - fu \quad (2)$$

$$\frac{\partial}{\partial y}(1) \Rightarrow \frac{D_h}{Dt}\left(\frac{\partial u}{\partial y}\right) = -\frac{1}{\rho}\frac{\partial^2 p_d}{\partial x \partial y} + f\frac{\partial v}{\partial y} + v\frac{\partial f}{\partial y} \quad (1)'$$

$$\frac{\partial}{\partial x}(2) \Rightarrow \frac{D_h}{Dt}\left(\frac{\partial v}{\partial x}\right) = -\frac{1}{\rho}\frac{\partial^2 p_d}{\partial x \partial y} - f\frac{\partial u}{\partial x} - u\frac{\partial f}{\partial x} \quad (2)'$$

$$(2)' - (1)' \Rightarrow \frac{D_h}{Dt}\left(\frac{\partial v}{\partial x} - \frac{\partial u}{\partial y}\right) = -f\left(\frac{\partial u}{\partial x} + \frac{\partial v}{\partial y}\right) - u\frac{\partial f}{\partial x} - v\frac{\partial f}{\partial y}$$

$$\therefore \frac{D_h \zeta}{Dt} + u\frac{\partial f}{\partial x} + v\frac{\partial f}{\partial y} = -f\left(\frac{\partial u}{\partial x} + \frac{\partial v}{\partial y}\right)$$

where we have assumed that the density ρ does not vary appreciably in the horizontal, which is consistent with the scaling we are using (Section 5.4.2). Now, since we also know that the Coriolis parameter does not change with time, we can write this equation more efficiently as

$$\frac{D_h}{Dt}(\zeta + f) = -f\left(\frac{\partial u}{\partial x} + \frac{\partial v}{\partial y}\right) \tag{7.7}$$

Thus, the synoptically scaled vorticity equation tells us that changes in time of relative vorticity are generated by the advection of relative and planetary vorticity, and by the convergence of planetary vorticity carrying air.

This conclusion suggests that if we want to understand the temporal evolution of a mid-latitude cyclone, we should use a theoretical model that allows convergence and divergence. Consider geostrophic flow: in this approximation the flow is non-divergent and hence the expression above for the change in total vorticity is identically zero – in this case total vorticity (but not relative vorticity alone) is conserved following the motion. More interesting is the case of quasi-geostrophic flow. In this case, the total flow is non-divergent, but the model allows divergence in the horizontal to be balanced by convergence in the vertical, and vice versa. By adopting the scaling leading to Equation (6.15), we can write Equation (7.7) as

$$\frac{D_g}{Dt}(\zeta_g + f) = -f\left(\frac{\partial u_a}{\partial x} + \frac{\partial v_a}{\partial y}\right) \tag{7.8}$$

which can also be written, using Equation (6.14), as

$$\frac{D_g}{Dt}(\zeta_g + f) = f\frac{\partial w}{\partial z}$$

Hence, this equation shows us that relative vorticity in a quasi-geostrophic situation will be generated by the advection of relative vorticity, computed geostrophically, and planetary vorticity, and by the convergence of planetary vorticity carrying air. By our use of Equation (6.14), we can understand this last term equivalently as the vertical stretching of planetary vorticity tubes.

Review questions

7.1 Consider a rectangular region of the atmosphere, with an east/west extent of 500 km and a north/south extent of 200 km. From the southern to the northern edge of this region the zonal wind increases from $20\,\mathrm{m\,s^{-1}}$ to $50\,\mathrm{m\,s^{-1}}$, and there is no meridional wind. Assume that the zonal wind only varies in the meridional direction.

(a) Calculate the circulation around this rectangular region.

(b) Does the sign of the circulation calculated in part (a) match the expected sense of rotation induced by this flow?

(c) Repeat parts (a) and (b) by assuming that the wind decreases from $50\,\mathrm{m\,s^{-1}}$ to $20\,\mathrm{m\,s^{-1}}$ from the southern to northern edge of this rectangular region.

(d) Calculate the relative vorticity for the flow in parts (a) and (c) using (i) the circulation calculated above, and (ii) Equation (7.3).

7.2 Heavy morning snow has just fallen on a plain east of Bismarck, North Dakota. Meanwhile, news radio from Pierre, South Dakota reports no snow in the state that day.

(a) Write at least one paragraph describing why and how a local thermally driven circulation is likely to develop during the afternoon. Support your description with a diagram. You can assume: (i) the sky clears; (ii) no further snow falls; and (iii) no large-scale pressure gradient.

(b) Calculate the rate of change of the circulation between the 1000 mb and 850 mb levels, and between a location 20 km north of the snow edge to a location 20 km south of the snow edge. Assume a temperature difference of 10 K between the two regions and an average temperature of 278 K. Estimate the wind speed 1 hour after the circulation commences.

7.3 In the late autumn, with the onset of longer nights and shorter days, Lake Superior in the northern United States remains relatively warm while the surrounding land begins to cool. The temperature difference between the land and adjacent lake is largest at night, and is confined to the lowest 50 hPa of the atmosphere.

(a) Calculate the wind speed between the land and the lake 2 hours after the horizontal temperature difference described above has been established, assuming that the temperature difference between 10 km inland to 10 km offshore is 5 K and that the average temperature is 278 K. You may assume that at sunset there is initially no temperature difference or horizontal pressure gradient between the land and the lake.

(b) Is this wind directed from the land to the lake or from the lake to the land? Why?

7.4 Calculate the relative vorticity at 500 mb over Amarillo, Texas, at 00 UTC 15 Feb 2003 (Figure 1.8) using the wind observations at this level from Albuquerque, New Mexico; Midland, Texas; Oklahoma City, Oklahoma; and Dodge City, Kansas.

7.5 Hurricane Katrina became the third strongest hurricane in the Atlantic Basin as it moved over the Gulf of Mexico in August 2005. Between 15 UTC 26 Aug and 15 UTC 28 Aug 2005 the hurricane intensified rapidly as the central pressure decreased from 981 hPa to 907 hPa.

(a) Calculate the relative vorticity of this hurricane at 15 UTC 26 Aug when the maximum winds were observed to be $36 \, m \, s^{-1}$ at a distance of 14 km from the center of the hurricane.

(b) Calculate the relative vorticity of this hurricane at 15 UTC 28 Aug when the maximum winds had increased to $77 \, m \, s^{-1}$ at a distance of 18.5 km from the center of the hurricane.

7.6 Construct a table similar to Table 7.1 for westerly flow over a mountain range in the Southern Hemisphere.

7.7 Construct a table similar to Table 7.1 for easterly flow over a mountain range in the Northern Hemisphere.

7.8 A westerly flow at 50°N latitude is approaching the Canadian Rocky Mountains, and will rise adiabatically over the mountains. As the flow approaches the mountains, it is bounded by two potential temperature surfaces: 296 K at a pressure of 850 hPa and 320 K at a pressure of 300 hPa. You may assume that frictional effects are negligible.

(a) If there is no meridional flow initially, and no horizontal shear in the zonal wind, what is the initial relative and absolute vorticity of the flow before it reaches the mountains?

(b) Calculate the relative and absolute vorticity of the flow at the crest of the mountains, assuming that the upper potential temperature surface is still at a pressure of 300 hPa, while the lower potential temperature surface has risen to 700 hPa, and the flow has been deflected 5° of latitude toward the south.

7.9 (a) What terms in the Navier–Stokes equations (4.19) have been neglected in the derivation of Equation (7.7)?

(b) Is it reasonable to neglect these terms for a synoptic scale mid-latitude weather system? Why or why not?

7.10 (a) Rewrite Equation (7.7) by expanding the material derivative in this equation.

(b) Describe what each term in this equation represents physically.

(c) Which terms in this expanded equation are equal to zero, and why?

7.11 (a) Calculate the relative vorticity at Oklahoma City, Oklahoma at 500 hPa at 00 UTC and 12 UTC 16 Feb 2003 based on the adjacent wind observations (use the storm of 2003 weather maps on the CD-ROM).

(b) Using the wind observations at 00 UTC calculate the local rate of change of the relative vorticity due to the convergence term in Equation (7.7).

(c) Based only on the change in relative vorticity due to convergence, what value of relative vorticity would you expect at 500 hPa over Oklahoma City at 12 UTC 16 Feb 2003?

(d) How does this compare to the value calculated from the 500 mb map at this time? Why would the value predicted based on the 00 UTC data differ from that calculated from the observations at 12 UTC?

7.12 (a) Rewrite Equation (7.8) by expanding the material derivative in this equation.

(b) Describe what each term in this equation represents physically.

(c) Which terms in this expanded equation are equal to zero, and why?

8 Simple wave motions

The dual concepts of vorticity and waves will provide us with the basis for understanding how mid-latitude cyclones work. Vorticity is important because, as we have seen, we can use it to create quantities that are conserved in some situations, and the interplay between changes in planetary vorticity and changes in relative vorticity can generate interesting motion. Waves are important because all weather systems can be understood physically as waves with particular wavelengths, excited by various forces acting on the atmosphere. This is why particular weather systems have characteristic scales – the scales are associated with the wavelengths and frequencies of the excited waves.

8.1 Properties of waves

In this section, we review some basic definitions in the description of waves. Such a disturbance can take the form of a traveling or *propagating* wave, or a *standing* wave. The simplest of such motions is the linear harmonic oscillator – for example, a pendulum (Figure 8.1).

In such a system the limits of oscillation are equally spaced around the equilibrium position. It satisfies the equation

$$\theta = \theta_0 \cos(\nu t - \alpha)$$

so that the motion can be completely described by knowing the *amplitude* θ_0, the *frequency* ν, and the *phase* $\phi = (\nu t - \alpha)$ of the pendulum swing. The constant α is called the phase constant.

The amplitude denotes the maximum deflection of the pendulum. The frequency tells us how many complete oscillations are performed by the pendulum in a given period of time. The unit of frequency is cycles per second, or hertz (Hz). The phase of a wave is determined from the location of the pendulum mass at an identified time $t = 0$. From a mathematical point of view, this is the value of θ as the cosine curve crosses the t axis. The amplitude and the phase of a particular oscillation depend on the initial position and initial speed, but the frequency depends only on the length of the string (a convenient attribute for early clock makers.)

Applied Atmospheric Dynamics Amanda H. Lynch, John J. Cassano
© 2006 John Wiley & Sons, Ltd

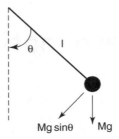

Figure 8.1 A pendulum of mass M is suspended on a frictionless, massless string of length l

In the atmosphere, wave motions are oscillations in field variables such as velocity and pressure. In this context, traveling or propagating waves are most important. Such waves transport energy, but not matter. Because the wave is traveling through the medium, we need to include position as well as time in our description of the disturbance. Most generally, this can be written

$$\theta = \theta_0 \cos(kx + ly + mz - \nu t - \alpha)$$

Now the phase $\phi = (kx + ly + mz - \nu t - \alpha)$ becomes a more complex concept, because we need to fix ourselves in time and in space to determine what part of the oscillation we are observing. The vector represented by $\vec{k} = (k, l, m)$ is known as the *wavenumber*, which is related to the wavelength λ by

$$\left|\vec{k}\right| = \frac{2\pi}{\lambda}$$

The *phase speed* of the wave, a scalar variable, is the rate at which the phase of the wave propagates in each of the three spatial dimensions, and is given by

$$c_x = \frac{\nu}{k} \quad c_y = \frac{\nu}{l} \quad c_z = \frac{\nu}{m}$$
$$c = \frac{\nu}{\left|\vec{k}\right|} \tag{8.1}$$

Since the phase speed depends on the wavenumber, waves of different wavelengths, starting from the same place, will propagate at different rates and hence will spread out, or disperse. Because of this property, the relationship between frequency and wavenumber for any given wave is called the *dispersion relation*. In the situation where $\nu \propto \left|\vec{k}\right|$, the phase speed is not a function of wavenumber, and the wave is called *non-dispersive*.

The rate at which the envelope of observable disturbance, and, with it, the energy of the waves, travels is a vector quantity called the *group velocity*. This means that when a disturbance is made up of several waves of different phases and amplitudes, we can determine a rate and direction of travel that are representative of the entire group of waves. The components of the group velocity are determined by

$$c_{gx} = \frac{\partial \nu}{\partial k} \quad c_{gy} = \frac{\partial \nu}{\partial l} \quad c_{gz} = \frac{\partial \nu}{\partial m} \quad (8.2)$$

Example Consider an atmospheric wave in which buoyancy is the restoring force. To simplify, assume the wave is in only two dimensions (x and z). The frequency of such a wave is given by

$$\nu = \frac{Nk}{|\vec{k}|} = \frac{Nk}{\sqrt{k^2 + m^2}}$$

where N is the Brunt–Väisälä frequency (Section 3.4). The phase speeds of this wave in the zonal and vertical directions are

$$c_x = \frac{\nu}{k} \qquad c_z = \frac{\nu}{m}$$
$$= \frac{N}{|\vec{k}|} \qquad = \frac{Nk}{|\vec{k}|m}$$

and the total phase speed in the direction of travel is

$$c = \frac{\nu}{|\vec{k}|} = \frac{Nk}{|\vec{k}|^2}$$

Note that the zonal and vertical phase speeds are *not* components of a vector whose magnitude is the total phase speed. However, the group velocity is a vector, whose components are given by

$$c_{gx} = \frac{\partial \nu}{\partial k} \qquad c_{gz} = \frac{\partial \nu}{\partial m}$$
$$= \frac{Nm^2}{|\vec{k}|^3} \qquad = \frac{-Nkm}{|\vec{k}|^3}$$

An example of just such a wave, known in atmospheric dynamics as a 'gravity wave', is shown on the CD-ROM and described in more detail in Chapter 13.

8.2 Perturbation analysis

Perturbation analysis is a mathematical tool that allows us to study waves in the atmosphere by making some rather strict assumptions to remove the nonlinearities in the chosen equations of motion (see Section 5.4.2). These assumptions are very different from, for example, the scale assumptions of geostrophic flow, since they are designed to reflect specifically the behavior of waves. In perturbation analysis, we consider that all variables are a sum of some (yet to be defined) basic state and some small departure from that state. The basic state and the departure can be functions of space and time, but an important requirement is that the basic state variables themselves satisfy the governing equations. A second important requirement is that the departures from the basic state, known as the perturbations, are sufficiently small that any products of perturbation terms can be neglected.

Example Consider the x component of the quasi-geostrophic horizontal momentum equation

$$\frac{D_g u_g}{Dt} + f u_a = 0$$

We define a basic state and a perturbation for each variable in the equation:

$$u_g(x, y, t) = \bar{u}_g(x, t) + u'_g(x, y, t)$$
$$u_a(x, y, t) = \bar{u}_a(x, t) + u'_a(x, y, t)$$
$$v_g(x, y, t) = \bar{v}_g(x, t) + v'_g(x, y, t)$$

where we have neglected variations in height since they do not appear in our equation of interest. Substituting into the equation gives

$$\frac{D_g}{Dt}(\bar{u}_g + u'_g) + f(\bar{u}_a + u'_a) = 0$$

$$\frac{\partial \bar{u}_g}{\partial t} + \frac{\partial u'_g}{\partial t} + \bar{u}_g \frac{\partial \bar{u}_g}{\partial x} + \bar{u}_g \frac{\partial u'_g}{\partial x} + u'_g \frac{\partial \bar{u}_g}{\partial x} + u'_g \frac{\partial u'_g}{\partial x} + \bar{v}_g \frac{\partial u'_g}{\partial y} + v'_g \frac{\partial u'_g}{\partial y} + f\bar{u}_a + f u'_a = 0$$

Omitting quantities in which products of perturbation terms appear yields

$$\frac{\partial \bar{u}_g}{\partial t} + \frac{\partial u'_g}{\partial t} + \bar{u}_g \frac{\partial \bar{u}_g}{\partial x} + \bar{u}_g \frac{\partial u'_g}{\partial x} + u'_g \frac{\partial \bar{u}_g}{\partial x} + \bar{v}_g \frac{\partial u'_g}{\partial y} + f\bar{u}_a + f u'_a = 0$$

In addition, we have defined that the basic state satisfies the governing equation alone, that is

$$\frac{\partial \bar{u}_g}{\partial t} + \bar{u}_g \frac{\partial \bar{u}_g}{\partial x} + f \bar{u}_a = 0$$

so we can derive the governing equation for the time evolution of the perturbation field:

$$\frac{\partial u'_g}{\partial t} + \bar{u}_g \frac{\partial u'_g}{\partial x} + u'_g \frac{\partial \bar{u}_g}{\partial x} + \bar{v}_g \frac{\partial u'_g}{\partial y} + f u'_a = 0$$

Note that this equation is linear.

Using this approach, we can create a linear version of the quasi-geostrophic vorticity equation (Equation (7.8)). We begin by assuming that the vertical velocity is identically zero, and the basic state is a westerly zonal flow $\vec{u} = (u, 0)$ which is constant, or at most a function of t:

$$\frac{D_g}{Dt}(\zeta_g + f) = f \frac{\partial w}{\partial z}$$

$$\frac{D_g \zeta_g}{Dt} + v_g \frac{\partial f}{\partial y} = 0$$

Such a system is non-divergent. Let us also assume that the Coriolis parameter varies linearly with latitude. This is known as the *beta plane approximation*, since we write f as

$$f = f_0 + \beta y$$

In general,

$$\beta = \frac{2\Omega \cos \phi_0}{a} \tag{8.3}$$

where ϕ_0 is the reference latitude, that is the latitude at which the plane is tangent to the Earth. When we center f_0 on the middle latitudes, β has a magnitude of around 10^{-11}, and the approximation is quite accurate for small departures from the reference latitude. Using the beta plane approximation, we can derive the perturbation form of the equation:

$$\frac{D_g \zeta_g}{Dt} + \beta v_g = 0$$

$$\Rightarrow \frac{\partial \bar{\zeta}_g}{\partial t} + \frac{\partial \zeta'_g}{\partial t} + \bar{u}_g \frac{\partial \bar{\zeta}_g}{\partial x} + \bar{u}_g \frac{\partial \zeta'_g}{\partial x} + u'_g \frac{\partial \bar{\zeta}_g}{\partial x} + u'_g \frac{\partial \zeta'_g}{\partial x} + v'_g \frac{\partial \bar{\zeta}_g}{\partial y} + v'_g \frac{\partial \zeta'_g}{\partial y} + \beta v'_g = 0$$

$$\frac{\partial \bar{\zeta}_g}{\partial t} + \frac{\partial \zeta'_g}{\partial t} + \bar{u}_g \frac{\partial \bar{\zeta}_g}{\partial x} + \bar{u}_g \frac{\partial \zeta'_g}{\partial x} + u'_g \frac{\partial \bar{\zeta}_g}{\partial x} + v'_g \frac{\partial \bar{\zeta}_g}{\partial y} + \beta v'_g = 0 \quad \text{omit perturbation products}$$

$$\frac{\partial \zeta'_g}{\partial t} + \bar{u}_g \frac{\partial \zeta'_g}{\partial x} + \beta v'_g = 0 \quad \text{zonal basic flow: } \bar{\zeta}_g = 0$$

We can find a wave-like solution to this equation by assuming a solution of the form

$$u'_g = -lA \cos(kx + ly - vt)$$

$$v'_g = kA \cos(kx + ly - vt)$$

We use this form for the perturbation wind field to ensure that the solution is non-divergent. By deriving an expression for ζ'_g and substituting the assumed solution into the simplified equation we can determine the dispersion relation for this wave:

$$\zeta'_g = \frac{\partial v'_g}{\partial x} - \frac{\partial u'_g}{\partial y} = -A(k^2 + l^2)\sin(kx + ly - \nu t)$$

$$\frac{\partial \zeta'_g}{\partial t} + \bar{u}_g \frac{\partial \zeta'_g}{\partial x} + \beta v'_g = 0$$

$$-\nu(-A(k^2 + l^2))\cos(kx + ly - \nu t) + k\bar{u}_g(-A(k^2 + l^2))\cos(kx + ly - \nu t)$$
$$+ \beta A k \cos(kx + ly - \nu t) = 0$$

$$\nu A(k^2 + l^2) - k\bar{u}_g A(k^2 + l^2) + \beta A k = 0$$

$$\Rightarrow \nu = \frac{-\beta k}{(k^2 + l^2)} + k\bar{u}_g \tag{8.4}$$

As we can see, this wave is certainly dispersive (that is, different wavelengths travel at different phase speeds). In fact, from the dispersion relation, we can see that the phase speed in the zonal direction is

$$c_x = \frac{\nu}{k}$$

$$\Rightarrow c_x - \bar{u}_g = -\frac{\beta}{\left|\vec{k}\right|^2} \tag{8.5}$$

Hence, we see that longer waves (with a smaller wavenumber) travel faster than shorter waves. Ignoring the contribution of the zonal wind, propagation is toward the west (that is, in the negative x direction). Equation (8.5) is known as *Rossby's formula*, named after Carl-Gustav Rossby, who substantially (though not solely) developed this analysis in a series of important papers in the late 1930s and early 1940s.

8.3 Planetary waves

The formulation we derived in Section 8.2 made some very strong assumptions in order to simplify the solution of the equations. Our assumption of non-divergence (equivalent to assuming barotropic flow) can only be justified if we are considering the scales of motion in the atmosphere that are much larger than mid-latitude synoptic systems. At such large scales, disturbances in the atmosphere are two to three orders of magnitude broader than they are deep, and the assumption of barotropy is not so problematic. Hence, it is appropriate to apply this equation to the largest scales of flow away from the surface, and in fact Rossby's formula describes very well the behavior of *planetary waves*, shown in Figure 8.2. These waves circle the globe in the middle latitudes, and can be seen in this map as peaks above 1550 m and

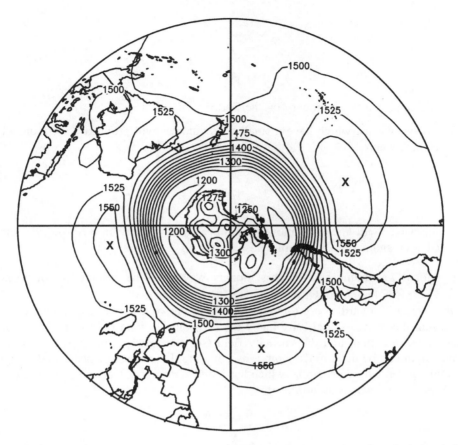

Figure 8.2 The 850 hPa geopotential height (m) over the Southern Hemisphere during January 2003. Peaks in the wavenumber 3 pattern are labeled with an 'X'. NCEP Reanalysis data provided by the NOAA-CIRES Climate Diagnostics Center, Boulder, Colorado, from its Web site at http://www.cdc.noaa.gov/

troughs below 1525 m in the geopotential height field. The figure shows roughly three peaks (indicated by an 'X'), and hence this example is called a 'wavenumber 3' planetary wave.

We see from Equation (8.5) that the planetary waves, also known as *Rossby waves*, are advected with the basic zonal flow. Because of this, we may consider three possible situations:

1. Shorter waves, which have large wavenumber, will propagate more slowly. If

$$\frac{\beta}{\left|\vec{k}\right|^2} < \bar{u}_g$$

the waves will appear to move slowly eastward.

2. Medium waves in which

$$\frac{\beta}{|\vec{k}|^2} = \bar{u}_g$$

will appear to be stationary on the Earth.

3. Longer waves, which have a small wavenumber, will propagate more quickly. Hence, when

$$\frac{\beta}{|\vec{k}|^2} > \bar{u}_g$$

the waves will appear to move slowly westward; that is, in a *retrograde* sense (that is, against the prevailing westerly flow.)

The critical wavenumber at which stationary waves will be observed is, of course, $\sqrt{\beta/\bar{u}_g}$. This is known as $\vec{k}_s$, the stationary wavenumber.

It is straightforward to use this relationship to derive the other aspects of interest regarding this wave. We already know the form of the wind field in the zonal and meridional direction, and the vorticity field (see the example in Section 8.2). The wavelength can be determined simply by counting the number of peaks and troughs around a given latitude ϕ circle, which has length $2\pi a \cos \phi$, where a is the radius of the Earth. For n waves at latitude $45°$,

$$\lambda = \frac{2\pi a \cos \phi}{n}$$

$$= \frac{\sqrt{2}\pi a}{n}$$

$$\Rightarrow k = \frac{2\pi}{\lambda} = \frac{\sqrt{2}n}{a}$$

$$c_x = \bar{u}_g - \frac{\beta}{|\vec{k}|^2}$$

$$= \bar{u}_g - \frac{\beta a^2}{2n^2}$$

Example Since for $\phi_0 = 45°$, $\beta = \sqrt{2}\Omega/a$, we can determine the period T for n waves around a latitude circle. Neglecting the zonal wind,

$$T = \left|\frac{\lambda}{c_x}\right|$$

$$= \frac{\sqrt{2}\pi a}{n} \times \frac{2n^2}{\beta a^2}$$

$$= \frac{\sqrt{2}\pi a}{n} \times \frac{2n^2 a}{\sqrt{2}\Omega a^2}$$

$$= \frac{2\pi n}{\Omega}$$

Since $\Omega = 2\pi/\text{day}$, the period is simply n days. Hence, for example, the wavenumber 3 pattern that we see in Figure 8.2 has a period of 3 days, a wavelength of

$$\lambda = \frac{\sqrt{2}\pi a}{n} = \frac{\sqrt{2}\pi \times 6.37 \times 10^6}{3} = 9.43 \times 10^6 \, \text{m}$$

and a phase speed, excluding the effects of a prevailing zonal flow, of

$$c_x = -\frac{\beta a^2}{2n^2}$$

$$= -\frac{\Omega a}{\sqrt{2}n^2}$$

$$= -\frac{7.292 \times 10^{-5} \times 6.37 \times 10^6}{\sqrt{2} \times 9}$$

$$= -36.5 \, \text{m s}^{-1}$$

This calculation also tells us that for a prevailing westerly zonal flow of $36.5 \, \text{m s}^{-1}$, or 70.9 kts, 3 is the stationary wavenumber.

The vorticity equation we used to derive this information about planetary scale waves was rendered geostrophic by our assumption of non-divergent flow. As a result, the restoring force for the wave can be understood by considering the force balance between Coriolis and pressure gradient forces expressed in the geostrophic wind relationship. Because the Coriolis force increases with latitude, so too does the pressure gradient, providing the restoring force for the wave motion. However, the conservation of potential vorticity (Section 7.3) gives a clearer demonstration of how planetary waves work. In fact, the example discussed in that section described a topographically forced Rossby wave.

8.3.1 Forcing of planetary waves

Although the planetary wave patterns take on a great variety of configurations, there is a tendency for the longer waves to occur in preferred geographic locations. Figure 8.3 shows the 500 hPa geopotential height *anomaly* (difference from the long-term mean) for the northern winters from 1968 to 1996. Using the anomaly pattern we can discern

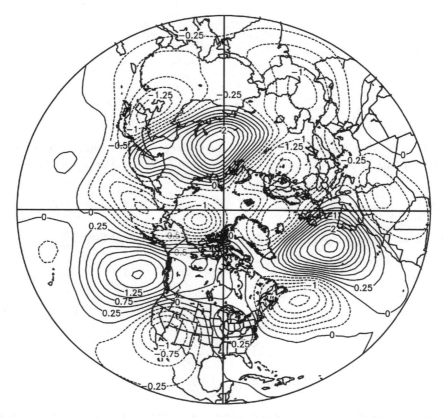

Figure 8.3 The 500 hPa geopotential height (m) anomaly for the winter season (DJF) over the Northern Hemisphere for the period 1968 to 1996. NCEP Reanalysis data provided by the NOAA-CIRES Climate Diagnostics Center, Boulder, Colorado, from its Web site at http://www.cdc.noaa.gov/

the preferred locations of peaks and troughs of the geopotential height and see that they are associated with meridionally oriented mountain ranges.

In the Northern Hemisphere winter, for example, a four-wave pattern is apparent, with troughs located in the North Atlantic, in northern␣Scandinavia southward into the Mediterranean Sea, the Aleutian region of the north Pacific, and in the vicinity of the Rocky Mountains in North America. The persistence of particular patterns seems to suggest that the forcing mechanisms for these waves also exhibit some kind of persistence and that the waves are quasi-stationary. There must also be a mechanism that allows variability. Of course, at any given moment the atmosphere will not look like this pattern, not just because of this variability, but because transient disturbances (weather) can be as large as the more slowly varying Rossby wave disturbance.

In Figure 8.3, we see that troughs are located preferentially on the lee side (to the east) of mountain ranges, with relatively higher pressure on the windward or upstream side. This can be interpreted as the atmosphere exerting an eastward force on the solid Earth, with the Earth exerting an equal and opposite force on the atmosphere.

In fact, forced waves are generated by three principal mechanisms. In addition to the topographic forcing that can be inferred from the basic patterns, Rossby waves are generated by thermal forcing due to longitudinal heating differences associated with the distribution of oceans and continents, and by forcing due to nonlinear interactions with smaller scale disturbances such as extra-tropical cyclones.

Consider now the vertical variation of geopotential height anomaly (Figure 8.4). The patterns vary little with height, although the wind does become stronger as one reaches the tropopause (usually round 150 hPa at middle latitudes.) This confirms that our use of the barotropic model is quite appropriate. The more interesting feature is that the pattern tilts westward with height. This is a reflection of the upward transport of energy and westward momentum, as successive layers of air exert a westward pressure force on the layer above.

The simplest approach to solve for a topographically forced Rossby wave is to assume a sinusoidal mountain and search for a stationary solution (that is, omit any

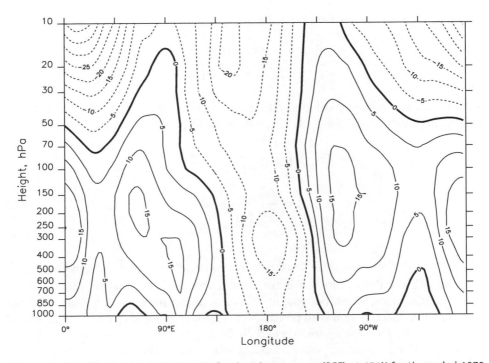

Figure 8.4 Geopotential height anomaly for the winter season (DJF) at 45°N for the period 1975 to 2005. NCEP Reanalysis data provided by the NOAA-CIRES Climate Diagnostics Center, Boulder, Colorado, from its Web site at http://www.cdc.noaa.gov/

variation with time). Drawing from the pioneering work of Charney and Eliassen (1949), we can return to Equation (7.8):

$$\frac{D_g}{Dt}(\zeta_g + f) = f\frac{\partial w}{\partial z}$$

In a barotropic fluid, the thermal wind vanishes, and hence the wind and the vorticity are not functions of z. This allows us to integrate the above equation from a lower boundary $h(x, y)$ to an upper boundary at the tropopause, which is assumed to be at constant height H. Then,

$$(H - h(x, y))\frac{D_g}{Dt}(\zeta_g + f) = f(w(H) - w(h)) = f\left(\left.\frac{Dz}{Dt}\right|_H - \left.\frac{Dz}{Dt}\right|_h\right)$$

$$\Rightarrow H\frac{D_g}{Dt}(\zeta_g + f) = -f\frac{Dh}{Dt} = -f\bar{u}_g\frac{\partial h}{\partial x}$$

if we assume that $h(x, y) \ll H$. Defining the lower boundary as a sinusoidal mountain,

$$h(x, y) = h_0 \sin(kx + ly)$$

we then assume a stationary wave solution:

$$u'_g = -lA\cos(kx + ly)$$
$$v'_g = kA\cos(kx + ly)$$

Substituting this assumed solution into the governing equation yields an expression for the amplitude of the wave, A, rather than the dispersion relation:

$$A\left(|\vec{k}|^2 - \frac{\beta}{\bar{u}_g}\right) = \frac{fh_0}{H}$$

$$A = \frac{fh_0}{H}\frac{1}{\left(|\vec{k}|^2 - k_s^2\right)} \tag{8.6}$$

We have in this solution an unrealistic case when $|\vec{k}| = k_s$ and the amplitude goes to infinity. Since this occurs at the zonal wind speed for which a free Rossby wave becomes stationary, it can be thought of as a resonant mode, which is an expected property of a linear system with no damping. Charney and Eliassen (1949) addressed this problem by introducing friction to their model. This removed the singularity and shifted the phase of the response so that it was more in line with the observations. As a result, they developed a model which they rightly recognized as having the potential to explain the winter mean circulation in the middle latitudes, and enable the evolution of the polar front (Section 1.4) and the cyclone tracks associated with it.

Review questions

8.1 Sound waves that are audible to the human ear have frequencies ranging from 20 to 20 000 s^{-1}. In air at a temperature of 20°C the phase speed of sound waves is 343 m s^{-1}. This phase speed does not vary with wavelength. Consider a sound wave that is propagating in time and in the x direction only.

(a) Calculate the wavenumber and wavelength for this range of frequencies.

(b) Calculate the group velocity of sound waves for the smallest and largest frequency audible waves. How does the group velocity compare to the phase speed of these waves?

(c) How long does it take sound waves with frequencies of 20 s^{-1} and 20 000 s^{-1} to travel 1 km?

(d) Are sound waves dispersive or non-dispersive?

8.2 An explosion generates sound waves with frequencies of 1000 and 2000 s^{-1}. These sound waves have $k = 2.92$ m^{-1} and 5.83 m^{-1} respectively. Based on this information, calculate the phase speed for these sound waves. Does this match the value given for the phase speed of sound waves in question 8.1?

8.3 The waves discussed in the example in Section 8.1 are known as internal gravity waves. As shown in this example, the frequency of these waves depends on the Brunt–Väisälä frequency and the horizontal and vertical wavenumber of the wave.

(a) For an atmosphere with $N = 0.023$ s^{-1} find the frequency of waves with: (i) $k = 3.14 \times 10^{-3}$ m^{-1} and m $= 4.58 \times 10^{-3}$ m^{-1}; (ii) $k = 6.28 \times 10^{-3}$ m^{-1} and m $= 5.74 \times 10^{-3}$ m^{-1}.

(b) What is the horizontal and vertical wavelength of these waves?

(c) Calculate the horizontal and vertical phase speed of these waves.

(d) Calculate the horizontal and vertical group velocity of these waves.

(e) Are these waves dispersive? Why or why not?

8.4 (a) The value of the Coriolis parameter can be estimated using the beta plane approximation (Equation (8.3)). Using a reference latitude of 45°, calculate values of the Coriolis parameter using the beta plane approximation at latitudes $\phi = 45°$, 44°, 40°, 30°, 25°, and 20°.

(b) Use Equation (5.5) to calculate the value of the Coriolis parameter at these latitudes.

(c) At which latitudes can the beta plane approximation be used with errors of less than (i) 5% and (ii) 10%?

8.5 Verify that the assumed solution for planetary waves, used to derive Rossby's formula, is indeed non-divergent.

8.6 Sketch the zonal and meridional wind fields and the vorticity field for a Rossby wave over one wavelength in the x direction.

8.7 (a) Calculate the zonal wavelength and wavenumber for planetary waves at a latitude of 40°N if there are a total of two, three, four, and five waves around the latitude circle.

(b) Calculate the zonal phase speed of these waves assuming that $\bar{u}_g = 0 \,\mathrm{m\,s^{-1}}$.

(c) Which wave has the largest zonal phase speed? Is this consistent with the discussion in Section 8.3?

(d) What is the period of these waves?

(e) What mean zonal wind speed is required to cause each of these waves to be stationary?

8.8 (a) Using the weather maps for the storm of 2003 on the CD-ROM, estimate the wavelength of the wave at 500 hPa at 00 UTC 16 Feb 2003.

(b) Calculate $c_x - \bar{u}_g$ for this wave.

(c) Estimate the phase speed of this wave using the 500 hPa weather maps at 00 UTC and 12 UTC 16 Feb 2003.

(d) What value of $\bar{u}_g$ is required for the wave to have this phase speed? Is your estimated value of $\bar{u}_g$ consistent with the observed zonal winds at this time?

8.9 (a) At a latitude of 40°S what is the wavenumber of a stationary planetary wave if $\bar{u}_g = 5, 10,$ and $25 \,\mathrm{m\,s^{-1}}$?

(b) What is the wavelength of each of the waves in part (a)?

(c) What is the direction of propagation for waves with $k < k_s$? What is the direction of propagation for waves with $k > k_s$?

8.10 (a) Estimate the horizontal (zonal and meridional) scale of the Colorado Rocky Mountains, Himalayas, Andes Mountains, and Southern Alps of New Zealand.

(b) Calculate the amplitude and perturbation zonal and meridional geostrophic winds for planetary waves forced by each of these mountain ranges for $\bar{u}_g = 20 \,\mathrm{m\,s^{-1}}$. (You will need to estimate the height of each mountain range and can assume that the tropopause is at an altitude of 11 km.)

9 Extra-tropical weather systems

An important feature of the mid-latitude flow in the middle to upper troposphere is the wave-like structure of westerly flow (Figure 9.1). Some of these waves, as we have seen, are Rossby waves. These may be close to stationary, moving very slowly eastward with the westerly flow, or long enough to *retrogress* (move westward). In addition, short waves are embedded in the Rossby waves. There can be anywhere between 6 and 18 short waves traversing a hemisphere at any one time, with wavelengths ranging from 4500 km down to 1500 km. As the short waves move through the long-wave pattern, the various waves interfere with one another, resulting in the complex flows that we observe in weather maps.

Extra-tropical synoptic scale weather systems are observed as closed circulations at the surface but these are simply the surface expression of short waves in the upper westerlies, first introduced in Section 1.3.3. Recall from Section 1.4 that typical extra-tropical cyclones are frontal cyclones; that is, they originate at the polar front, a highly baroclinic zone. Hence, the behavior of the short waves cannot assume the barotropic model as was appropriate for long waves. While this results in a more complex mathematical treatment, there are aspects of these short waves that we can consider using highly simplified models.

9.1 Fronts

9.1.1 Margules' model

The simplest model representing a front as a discontinuity in air properties is Margules' model, first published in 1906 by the Austrian meteorologist Max Margules. In this model, a cold front is idealized as a tilted plane separating two homogeneous, geostrophic flows with different, but uniform, temperatures and densities. The effects of friction near the surface are explicitly excluded. To develop this model, we choose a coordinate system in which the x axis is normal to the front and the y axis is parallel to the front (Figure 9.2). The front is stationary and infinitely thin, and we assume that the across-front changes in temperature and density are relatively small (the Boussinesq approximation).

Applied Atmospheric Dynamics   Amanda H. Lynch, John J. Cassano
© 2006 John Wiley & Sons, Ltd

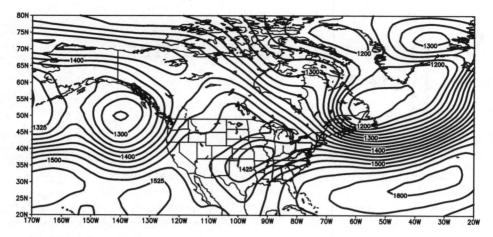

Figure 9.1 The 850 hPa geopotential height on 15 February 2003. NCEP Reanalysis data provided by the NOAA-CIRES Climate Diagnostics Center, Boulder, Colorado, from its Web site at http://www.cdc.noaa.gov/

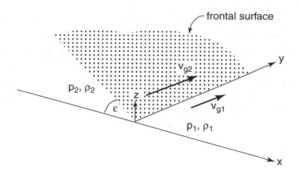

Figure 9.2 Idealized depiction of a frontal discontinuity

In such a model, the only flow is along the front, in the y direction, since an across-front wind would imply that the front was moving. Following Smith (1990), we define the pressure, density, and along-front wind in the warm air mass at p_1, ρ_1, and v_{g1}, and similarly for the cold air mass p_2, ρ_2, and v_{g2}. Because we are making the Boussinesq approximation, density differences across the front are important only inasmuch as they create buoyancy forces, and hence we can use the vertical momentum equation (6.10), where we have defined the reference density to be ρ_2:

$$\frac{1}{\rho_2}\frac{\partial p}{\partial z} + \frac{(\rho_2 - \rho_1)}{\rho_2}g = 0$$

Using this equation we can integrate from the ground to the height of the frontal surface $h(x)$, and determine how this varies by differentiating with respect to x, yielding

$$\frac{1}{\rho_2}\left(\frac{\partial p_1}{\partial x} - \frac{\partial p_2}{\partial x}\right) = -\frac{\rho_2 - \rho_1}{\rho_2} g \frac{\partial h}{\partial x} \tag{9.1}$$

Geometry tells us that $h(x) = -x \tan \varepsilon$ and hence we can write

$$\frac{1}{\rho_2}\left(\frac{\partial p_1}{\partial x} - \frac{\partial p_2}{\partial x}\right) = -\frac{\rho_2 - \rho_1}{\rho_2} g \tan \varepsilon$$

To determine the along-front wind, we simply apply the geostrophic relationship

$$f v_i = \frac{1}{\rho_i} \frac{\partial p_i}{\partial x}; i = 1, 2$$

to get

$$f\left(v_2 - \frac{\rho_1}{\rho_2} v_1\right) = -\frac{\rho_2 - \rho_1}{\rho_2} g \tan \varepsilon \tag{9.2}$$

which is *Margules' formula* for relating frontal slope to the along-front wind differential. While this is a highly simplified model, several practical conclusions may be drawn from it. For example, Equation (9.2) represents a balance between buoyancy forces and net Coriolis forces important for maintaining the stationary front. This balance is indicative of the dynamical processes in a real front. Also, the equation requires a cyclonic change in wind across the front. For example, in the northern hemisphere where $f > 0$, the along-front wind differential will be positive. Rarely, horizontal acceleration in a real front may be sufficiently large that the cyclonic wind shift at the front is suppressed, but more generally, this observation applies.

Example Figure 9.3 shows our case study cyclone as it reached Missouri on 15 February 2003. Do the weather reports from Oklahoma City and Little Rock, situated roughly the same distance from the front, on either side of it, indicate a cyclonic change in flow?

The station model for Oklahoma City tells us that the pressure was 1010.4 hPa, the temperature was 42°F (6°C), and the dew point was 36°F (2°C). The wind was 15 kts, from the NNW. This is clearly in the cold dry air mass, as would be expected. Little Rock reported a pressure of 1007.0 hPa, a temperature of 63°F (17°C), and a dew point of 61°F (16°C) and fog – a warm moist air mass in comparison. The wind was 10 kts, from the south. The report from Dallas, just ahead of the cold front, shows saturated air that has turned SW. Hence, we have a cyclonic (counterclockwise in the Northern Hemisphere) change in the wind reports.

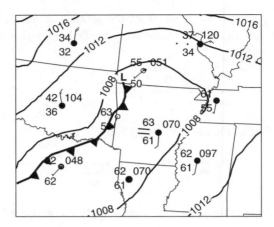

Figure 9.3 Sea level pressure and station reports over Arkansas and surrounding states on 15 February 2003 at 12 UTC. The warm front is omitted for clarity

Nevertheless, the shortcomings of this model are severe. The model does not take account of the vertical variations in temperature and density observed in the real atmosphere. More importantly, this model cannot represent a front which is moving normal to the frontal plane. This is because a simple coordinate transformation to an inertial frame moving with the speed of the front would introduce an additional Coriolis force in the along-front direction. The geostrophic approximation demands that this be balanced by a pressure gradient force, but Smith and Reeder (1988) show that this leads to an inconsistency, as follows.

Margules' model requires that we satisfy the requirement that the across-front wind component is continuous, that is

$$u_1 = u_2 \qquad (9.3)$$

and also the requirement that the pressure across the front be continuous. This second requirement has a corollary that the mass flux across the front be continuous (since pressure is an expression of the movement of mass), that is

$$\rho_1 u_1 = \rho_2 u_2 \qquad (9.4)$$

Since $\rho_1 \neq \rho_2$ we cannot simultaneously satisfy Equations (9.3) and (9.4) unless there is vertical motion (which violates the assumption of geostrophy) *or* $u_1 = u_2 = 0$. Hence, like the geostrophic model itself, Margules' model should be considered to be a useful conceptual 'snapshot' of a front rather than a truly descriptive model.

9.2 Frontal cyclones

The description of the common structure and dynamics of frontal cyclones (Section 1.4) is drawn from the remarkable synthesis of Bjerknes and Solberg, who drew together the many disparate observational and theoretical studies into a single conceptual model, shown in Figure 1.10. The basic principle developed in this work was that every cyclone is characterized by two fronts – the warm front in advance of the cyclone center (known earlier as the 'steering line') and the cold front behind the cyclone center (known earlier as the 'squall line'). The model was further developed by Bjerknes and Solberg to explain the formation and development of cyclones, and how the frontal structure is created (Figure 1.9). This process indicates that the nascent cyclone originates as a small wave whose amplitude then gradually increases through some instability mechanism. It is this process of *cyclogenesis* and the associated decay of cyclones, or *cyclolysis*, that will be discussed now.

9.2.1 Cyclogenesis

If a surface cyclone is to form or deepen, there must be net divergence of the mass of air in the region above the cyclone, for only then will the surface pressure decrease. As we saw in Section 1.4, a cyclone develops from a small wave in the polar front, which generates a region of relatively lower pressure at the surface. Due to the effects of friction near the ground disrupting the pure geostrophic balance, the air will spiral in toward the center of low pressure. In a deepening low, the isallobaric wind (Section 6.5.1) will also contribute significantly to the lower level convergence. This convergence adds air to the column and hence causes the surface pressure to increase. So, for surface pressure to continue to decrease, air must be removed from the column above the center of the cyclone faster than it is spiraling in at the surface. That is, the upper level divergence must outpace the lower level convergence.

We can explain why Rossby waves in the middle and upper troposphere generate convergence and divergence by representing the wind field associated with the wave to be the gradient wind, ignoring changes with time and frictional effects. By using the gradient wind model, we are allowing an ageostrophic component of the model, and hence we are allowing convergence and vertical motion (Equation (6.14)). Recall from Section 6.2.4 that in cyclones, the gradient wind is less than the geostrophic wind and in anticyclones the gradient wind is greater than the geostrophic wind. Hence, as air moves through such a wave, parallel to the pressure contours (Figure 9.4), it attains its highest speed in the ridges and lowest speed in the troughs, since the true wind is most accurately represented by the gradient wind. Thus, the wind speed must decelerate as the air moves from a ridge to a trough, causing a convergence of mass. Similarly, as we move from the low-speed trough to the high-speed ridge, acceleration and divergence must occur. And so these are the ideal conditions for what we might expect to see in a surface development – a region of upper level divergence occurring to the east of an upper level trough, allowing surface pressures to decrease, resulting in lower level convergence.

However, in order for cyclogenesis and cyclone intensification to occur, the upper level divergence must exceed the lower level convergence. How do we determine this? First, recall that our model suggests that cyclones form on the polar front. Due to thermal wind requirements, the strong horizontal temperature gradients at the front will lead to strong geostrophic winds aloft. These persistent strong winds

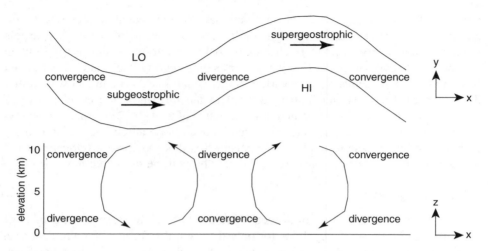

Figure 9.4 Lines of geopotential height are related to the gradient wind and the divergence field in the horizontal (top) and the vertical (bottom)

are called the polar jet stream. The polar jet stream circles the globe in the middle latitudes, experiencing considerable variations in latitude in response to the long waves and short waves propagating along the polar frontal zone. In addition, the jet stream experiences considerable variations in magnitude along its path – an example is shown on the CD-ROM. In fact, the jet speed can vary by a factor of 2 or even 3 – the zones of highest polar jet stream speeds are called *jet maxima* or jet streaks. Jet maxima are associated with short waves in the westerlies and can be thought of conceptually using the same gradient wind argument made above. Both the short waves and the accompanying jet maxima generally propagate eastward through the relatively slow moving Rossby waves.

What is generally observed is that cyclones form in specific regions of jet maxima, either on the left side of the jet exit or on the right side of the jet entrance. For example, the development of the offshore low on 16 and 17 February 2003 (Section 1.4.3) takes place in a location corresponding to the right side of the jet entrance, and the maximum is located over the North Atlantic (Figure 9.5).

So what is it about the jet maximum that causes divergence in the first place, and how does that process continue to intensify? Recall the example in Section 6.5.1 in which we found that, when a changing pressure field in space or in time creates an imbalance of forces, a cross-isobaric flow is generated. This cross-isobaric flow in turn creates regions of convergence and divergence and, as predicted by Equation (6.14), vertical motion. Let us apply this concept to the idealized Northern Hemisphere jet maximum depicted in Figure 9.6.

If we neglect changes in time and assume the geostrophic flow is approximately zonal, we can apply a simplified form of Equation (6.17):

$$v_a = \frac{1}{f} u_g \frac{\partial u_g}{\partial x}$$

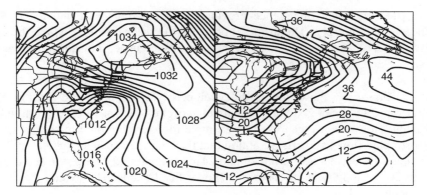

Figure 9.5 Sea level pressure (hPa, left) and 500 hPa wind shown as contours of isotachs (lines of constant wind speed) with some indicative direction vectors (m s^{-1}, right) on 17 February 2003 (both fields slightly smoothed), showing the location of the jet maximum in the north Atlantic, and the split jet over North America

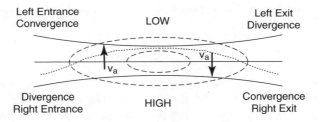

Figure 9.6 Horizontal view of the isobars (solid lines) and isotachs (dashed lines) with a typical parcel trajectory (dotted line) in an idealized jet maximum. Jet flow is in the x direction. The four quadrants of the jet are labeled

As in the earlier example from Section 6.5.1, the region of confluence at the jet entrance (as the isobars come together) results in u_g increasing as the flow enters the jet streak. Since $\partial u_g/\partial x > 0$ in the entrance region of the jet and all other terms in the equation above are positive in the Northern Hemisphere, this results in $v_a > 0$ (an ageostrophic flow in the positive y direction from south to north). Since this wind is toward low pressure, the flow extracts work from the pressure gradient and accelerates. In the jet exit, a region of diffluence (where isobars are spreading apart), $\partial u_g/\partial x < 0$, and a southward ageostrophic flow is generated – this flow is toward high pressure and decelerates. The cross-isobaric flow creates regions of convergence and divergence on either side of the jet exit and entrance, as shown, which in turn give rise to vertical motion. In the regions of upper level divergence, and thus ascending air, high-level cirrus cloud is often observed. In the regions of convergence, and descending air, any existing cloud may be expected to clear. In satellite pictures, the position of the jet is often indicated by a streak of high-level cloud, frequently with a sharp boundary along the jet axis.

Thus, the additional divergence provided on the left jet exit and the right jet entrance (in the Northern Hemisphere) provides the mechanism by which the upper level divergence can outpace the lower level convergence, and allow cyclones to intensify.

Example Figure 9.5 shows the newly developed cyclone on the east coast of North America on 17 February 2003. It is just about to occlude, and at 12 UTC it was already weakening. However, we can use this example to determine the typical magnitudes of ageostrophic flow and vertical motion that may occur in such a cyclone. In this jet maximum, air is accelerating from around $24\,\mathrm{m\,s^{-1}}$ to around $44\,\mathrm{m\,s^{-1}}$ over a distance of about $20°$ of longitude, or about 1500 km at this latitude. This would generate an ageostrophic wind of

$$v_a = \frac{1}{f} u_g \frac{\partial u_g}{\partial x} \approx \frac{1}{10^{-4}} \left(\frac{24+44}{2}\right) \left(\frac{44-24}{1500 \times 10^3}\right) \approx 4.5\,\mathrm{m\,s^{-1}}$$

Assuming that the air 500 km on either side of the jet is undisturbed, and the velocity maximum is around 2 km below the tropopause, the expected vertical motion would be

$$\frac{\partial w}{\partial z} = -\frac{\partial v_a}{\partial y} \Rightarrow w \approx 2 \times 10^3 \times \frac{4.5}{500 \times 10^3} \approx 2\,\mathrm{cm\,s^{-1}}$$

Such a vertical displacement would be sufficient to cause cloud formation in the ascending air. In this example, the development was aided by the warmth of the Gulf Stream flowing northward along the Atlantic coast driving vertical motion from below.

9.2.2 Sutcliffe Development Theory

Reginald C. Sutcliffe showed in 1938 that one can create a prognostic equation for cyclogenesis by using the vorticity equation. This approach is more accurate than trying to predict the net column divergence, because the integrated column divergence is a small residual of the often large contributions at various levels. This latter approach was used by Lewis F. Richardson in his famous numerical weather prediction experiment which he published in 1922 – in a 6 hour hand-calculated forecast still regarded as 'heroic', he predicted a pressure increase of 145 hPa for the area around Munich, Germany on 10 May 1910. Richardson himself identified the large apparent convergence of wind to be the source of the error. Sutcliffe's method using vorticity is a significant improvement on this approach.

Figure 9.7 shows an idealization of the vertical distribution of horizontal divergence and vertical motion for a mid-latitude cyclone. There is not an exact correspondence, level by level, between convergence and ascent. Ascent continues, for example, even through the lower levels of divergent flow. It is also clear that the curvature of the vertical velocity profile is consistently negative throughout the depth of the

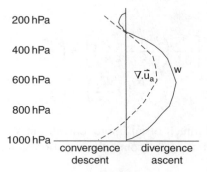

Figure 9.7 Typical patterns of divergence (dashed line) and ascent (solid line) in a mid-latitude cyclone

troposphere – our initial aim in developing this theory is to determine a diagnostic formula for this curvature. We can do so using the quasi-geostrophic vorticity equation (Equation (7.8)):

$$\frac{D_g}{Dt}(\zeta_g + f) = f\frac{\partial w}{\partial z}$$

$$\frac{\partial \zeta_g}{\partial t} + u_g\frac{\partial \zeta_g}{\partial x} + v_g\frac{\partial \zeta_g}{\partial y} + v_g\frac{\partial f}{\partial y} = f\frac{\partial w}{\partial z}$$

$$f\frac{\partial}{\partial z} \Rightarrow \frac{\partial}{\partial t}\left(f\frac{\partial \zeta_g}{\partial z}\right) + fu_g\frac{\partial^2 \zeta_g}{\partial x \partial z} + fv_g\frac{\partial^2 \zeta_g}{\partial y \partial z} + f\frac{\partial u_g}{\partial z}\frac{\partial \zeta_g}{\partial x} + f\frac{\partial v_g}{\partial z}\frac{\partial \zeta_g}{\partial y} + f\beta\frac{\partial v_g}{\partial z} = f^2\frac{\partial^2 w}{\partial z^2}$$

We must eliminate the time derivative from the left hand side of the equation to turn this from a prognostic into a diagnostic relation. To do this, we relate the hydrostatic equation in Boussinesq form (Equation (6.10)) to vorticity using Equation (5.11):

$$\zeta_g = \frac{\partial v_g}{\partial x} - \frac{\partial u_g}{\partial y} = \frac{1}{f\rho_{00}}\left(\frac{\partial^2 p}{\partial x^2} + \frac{\partial^2 p}{\partial y^2}\right)$$

Then by applying the operator

$$\left(\frac{\partial^2}{\partial x^2} + \frac{\partial^2}{\partial y^2}\right)\frac{\partial}{\partial t}$$

to

$$\frac{1}{\rho_{00}}\frac{\partial p_d}{\partial z} = -\frac{(\rho - \rho_0)}{\rho_{00}}g$$

we get

$$\frac{\partial}{\partial t}\frac{\partial}{\partial z}\left[\frac{1}{\rho_{00}}\left(\frac{\partial^2 p}{\partial x^2}+\frac{\partial^2 p}{\partial y^2}\right)\right]=\left(\frac{\partial^2}{\partial x^2}+\frac{\partial^2}{\partial y^2}\right)\frac{\partial \sigma}{\partial t}$$

$$\frac{\partial}{\partial t}\left(f\frac{\partial \zeta_g}{\partial z}\right)=\left(\frac{\partial^2}{\partial x^2}+\frac{\partial^2}{\partial y^2}\right)\frac{\partial \sigma}{\partial t}$$

This may not seem like much of an improvement, but we can then eliminate time from the buoyancy term using the continuity equation as follows:

$$\frac{D\rho}{Dt}=0$$

$$\frac{D}{Dt}(\rho_0+\rho-\rho_0)=0\times -\frac{g}{\rho_{00}}$$

$$\frac{D\sigma}{Dt}-\frac{g}{\rho_{00}}\frac{D\rho_0}{Dt}=0$$

$$\frac{D\sigma}{Dt}-\frac{g}{\rho_{00}}w\frac{d\rho_0}{dz}=0$$

We then define a buoyancy frequency $N_d^2=-gd\rho_0/gd\rho_{00}\partial z$ and employ quasi-geostrophic scaling to omit the advection of the buoyancy term by vertical motion and by the ageostrophic wind:

$$\frac{\partial \sigma}{\partial t}+u_g\frac{\partial \sigma}{\partial x}+v_g\frac{\partial \sigma}{\partial y}+wN_d^2=0$$

Using this, we can write a (rather complex) diagnostic equation for the curvature of the vertical velocity:

$$\left(\frac{\partial^2}{\partial x^2}+\frac{\partial^2}{\partial y^2}\right)\underbrace{\left(-u_g\frac{\partial \sigma}{\partial x}-v_g\frac{\partial \sigma}{\partial y}-wN_d^2\right)}_{(1)\quad\quad(2)}+\underbrace{fu_g\frac{\partial^2 \zeta_g}{\partial x\partial z}}_{(3)}+\underbrace{fv_g\frac{\partial^2 \zeta_g}{\partial y\partial z}}_{}+\underbrace{f\frac{\partial u_g}{\partial z}\frac{\partial \zeta_g}{\partial x}}_{(4)}$$

$$+f\frac{\partial v_g}{\partial z}\frac{\partial \zeta_g}{\partial y}+f\beta\frac{\partial v_g}{\partial z}=f^2\frac{\partial^2 w}{\partial z^2}$$
$$\quad\quad(5)\quad\quad\quad(6)$$

In Sutcliffe's derivation the adiabatic buoyancy term (2) was neglected, and he also determined that terms (1) and (3) cancel in part, leaving a quantity known as the deformation, which is small, and a term identical to term (4). By making these simplifications, the complex equation above reduces to a much simpler one:

$$2f\frac{\partial u_g}{\partial z}\frac{\partial \zeta_g}{\partial x}+2f\frac{\partial v_g}{\partial z}\frac{\partial \zeta_g}{\partial y}+f\beta\frac{\partial v_g}{\partial z}=f^2\frac{\partial^2 w}{\partial z^2}$$

Now, it is possible to integrate this equation from the surface to some height H. By choosing H to be a height at which the divergence is negligible, this can be written

$$f\frac{\partial w}{\partial z}\bigg|_0^H = -\int_0^H \left[2\frac{\partial u_g}{\partial z}\frac{\partial \zeta_g}{\partial x} + 2\frac{\partial v_g}{\partial z}\frac{\partial \zeta_g}{\partial y} + \beta\frac{\partial v_g}{\partial z}\right] dz \tag{9.5}$$

Thus, we have derived a technique for diagnosing the vertical velocity based on the vorticity. This relationship can also help to elucidate the mechanisms by which vorticity changes with time, in the quasi-geostrophic framework. If we can predict zones of increasing vorticity, these will correspond to likely cyclogenesis and intensification. Substituting Equation (9.5) back into Equation (7.8),

$$\frac{D_g}{Dt}(\zeta_g + f) = -\int_0^H \left[2\frac{\partial u_g}{\partial z}\frac{\partial \zeta_g}{\partial x} + 2\frac{\partial v_g}{\partial z}\frac{\partial \zeta_g}{\partial y} + \beta\frac{\partial v_g}{\partial z}\right] dz$$

$$\frac{\partial \zeta_g}{\partial t} = -u_g \frac{\partial \zeta_g}{\partial x} - v_g \frac{\partial \zeta_g}{\partial y} - \beta v_g - \int_0^H \left[2\frac{\partial u_g}{\partial z}\frac{\partial \zeta_g}{\partial x} + 2\frac{\partial v_g}{\partial z}\frac{\partial \zeta_g}{\partial y} + \beta\frac{\partial v_g}{\partial z}\right] dz \tag{9.6}$$

To simplify this further, let us assume that all quantities vary linearly with height. This constant vertical gradient is associated with the thermal wind variation and hence we will use the subscript T. Then,

$$\frac{\partial u_g}{\partial z} = \frac{u_T}{H} \quad \frac{\partial v_g}{\partial z} = \frac{v_T}{H} \quad \zeta_g = \zeta_0 + \zeta_T \frac{z}{H}$$

We can substitute these into Equation (9.6) to create an approximate prognostic equation for the surface vorticity:

$$\frac{\partial \zeta_{g0}}{\partial t} = \underbrace{-(u_{g0} + 2u_T)\frac{\partial \zeta_{g0}}{\partial x} - (v_{g0} + 2v_T)\frac{\partial \zeta_{g0}}{\partial y}}_{(1)} \underbrace{- u_T \frac{\partial \zeta_T}{\partial x} - v_T \frac{\partial \zeta_T}{\partial y}}_{(2)} \underbrace{- \beta(v_{g0} + v_T)}_{(3)} \tag{9.7}$$

This is Sutcliffe's development equation.

Despite the considerable simplifications and approximations that have been made, Sutcliffe's formulation, like Margules' model, yields important physical insights. Let us consider each of the terms on the right hand side of this equation.

At the center of a surface low-pressure system, the geostrophic wind speed will be much smaller than the wind shear, which is represented by u_T. Hence at this location, the advection term (1) can be written

$$-2u_T \frac{\partial \zeta_{g0}}{\partial x} - 2v_T \frac{\partial \zeta_{g0}}{\partial y}$$

This suggests that the low-pressure center will propagate in the direction of the wind shear, with a speed proportional to that wind shear. This is known as the *thermal steering principle.* Since we have assumed that $u_{g0} \approx 0$, then $u_T \approx u_{gH}$. This is why the lines of constant thickness at upper levels are often broadly similar to lines of geopotential height. It is often assumed that the level of zero divergence, H, is around the 500 hPa level. If this is approximately true, then the low-pressure system will tend to move in the direction of the 500 hPa wind. It is important to realize that the system moves through a process of development, not simply translation.

The 'thermal vorticity advection' term (2) is the primary contribution to the development of mid-latitude cyclones. Consider the short-wave pattern shown in Figure 9.8, represented by a wave in the thickness field. In such a diagram, the wind flowing along the wave is the wind shear. If we assume the wave is sinusoidal in form, the maximum gradient in ζ_T is one-quarter wavelength east of the thermal trough, and hence we can expect that term (2) is likely to be a maximum here also. This location is referred to as a *positive vorticity advection maximum*. Similarly, one-quarter wavelength east of the thermal ridge, we are likely to find a region of preferential anticyclonic vorticity generation.

The final term in the equation is simply the contribution of the planetary vorticity. Term (3) shows that cyclonic (that is, positive) vorticity is generated if the meridional wind is equatorward. The term is a small contributor to the overall development formula.

Used with care, Sutcliffe's development equation can be, and has proved to be, a very useful guide in weather forecasting. As with any simplified conceptual model, we must take account of its limitations when applying it. First, it neglects the effects of both adiabatic and diabatic heating. These often make important contributions to the overall evolution of the system. Indeed, the omission of the adiabatic term causes the speed of propagation to be overestimated by a factor of 2. Second, the formulation gives an estimate of tendencies at a given moment, or in practical terms, for a few hours. Thus, we would not expect Sutcliffe's equation to be a substitute for a good numerical weather prediction model. Finally, because of the assumptions made, the model fails badly when the shear is small. This problem is of particular

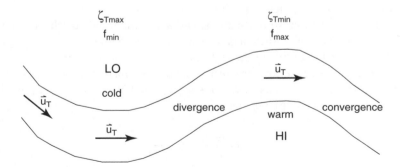

Figure 9.8 Lines of atmospheric thickness are related to the wind shear, the 'thermal vorticity', and the divergence field

concern when a system undergoes occlusion and is then isolated from the polar front and the associated strong shear.

9.2.3 Cyclolysis

The formation of an occluded front is often the first indication that a cyclone is about to dissipate. However, the process of occlusion is still a subject of active research. Even in the initial development of the theory, revisions occurred: originally Bjerknes believed that the occlusion, where the cold front catches up to the warm front, would initially occur away from the low center, but this view was probably influenced by observations made in Scandinavia at the time which were influenced by the local orography. Later work done in the United States by F. W. Reichelderfer suggested that the warm sector closes from the low center outward. Further, some scientists suggested that the 'neutral occlusion' suggested in the Norwegian model, in which thermal gradients disappeared at all levels in the atmosphere, would be unusual. This idea was supplanted with the ideas of *warm occlusions*, in which the polar air behind the cold front is warmer than the polar air ahead of the warm front, and *cold occlusions*, in which the reverse was true. It was suggested that cold occlusions would be more commonly observed, since air behind the advancing cold front was less likely to have been modified from cold stable polar conditions than air ahead of the advancing warm front. The concept was further developed by Tor Bergeron in 1937 to include the idea of a *bent back occlusion*, which occurs in the presence of secondary developments. In 1969, R. K. Anderson introduced the idea of *instant occlusions*, in which an occluded structure is created without the classic frontal 'catch-up' process. This was followed by a period in which observational studies consistently failed to verify the occlusion process, and in particular, the surface signals of occlusion.

So, the theory of occlusion and the Norwegian model in general was challenged for a considerable period of time. Key researchers suggested in the 1980s and early 1990s that the process did not occur. New conceptual models were developed. More recently, though, a combination of intense observational studies and model simulations have confirmed that the classical occlusion process does occur, though not in all cases (some examples are shown on the CD-ROM). Even now, it is frequently observed that a cyclone that appears to follow the classic life cycle will continue to intensify after occlusion appears to have taken place. So, refinement of this conceptual model continues.

9.3 Baroclinic instability

So far, we have been able to describe the basic ideas of Rossby waves, the polar front and the polar jet, short waves and jet maxima, and how all of these factors combine to form, intensify, and steer surface low-pressure systems. The missing piece from this puzzle is: what causes the development of short waves in the vicinity of jet maxima?

It has been observed that flows in which strong velocity shears occur are unstable with respect to small perturbations. That is, a small perturbation will grow rapidly. The role of a growing disturbance in any system is to bring the system back to a stable state. It follows that unstable conditions cannot persist in the atmosphere for any appreciable length of time. The transition from unstable to stable states involves a reduction of the potential energy, and all systems left to themselves will try to avoid instability and obtain a minimum of potential energy.

In this context, most synoptic scale systems in middle latitudes appear to develop as a result of this instability of the jet stream flow, which is closely associated through the thermal wind relationship with strong horizontal temperature gradients. A horizontal temperature gradient at the surface is reflected in the atmosphere above as slanting potential temperature surfaces whose height increases as one moves poleward. Then, the source of kinetic energy for the growing disturbances is extracted from the potential energy of the potential temperature field. This can occur when there is a small transverse displacement of air along a path that is less steep than the slope of the potential temperature surfaces. In this case, effectively colder air is lowered, or warmer air is raised: this is an energy releasing process and so allows the growth of the disturbance and reduces the potential energy of the atmospheric state. Because of the relationship to the temperature structure, this process is called *baroclinic instability*. A mathematical treatment of this process is beyond the scope of this book, and so we refer the reader to, for example, J. R. Holton's *An Introduction to Dynamic Meteorology* for more details.

Review questions

9.1 Using the surface weather observations in Oklahoma City, Oklahoma and Dodge City, Kansas estimate the slope of the cold front for the storm of 2003 shown on the surface weather map in Figure 1.7. What is the height of the cold front above the ground at Dodge City?

9.2 Consider a Northern Hemisphere cold front, oriented in the meridional direction, moving from west to east. The geostrophic wind in the warm air mass is from the south-west with a speed of $7.1 \, m \, s^{-1}$, while the geostrophic wind in the cold air mass is from the north-west. If there is no confluence or diffluence, what is the magnitude of the geostrophic wind speed in the cold air mass? Find the distance from the surface front at which the frontal plane is 1 km high, if the temperature difference is $10°C$, the average temperature is $15°C$, and the latitude is $45°N$. You may assume a pressure of 1000 hPa to calculate the density on the warm and cold sides of the front.

9.3 Consider a wave at the 500 hPa level that consists of a series of troughs and ridges, with radius of curvature of 500 km and a horizontal pressure gradient with a constant magnitude of $5 \times 10^{-4} \, Pa \, m^{-1}$. If the regions of cyclonic and anticyclonic curvature are separated by a region that is 200 km long where the height contours are straight and parallel, calculate the divergence across this region as the flow moves from a trough to a ridge and from a ridge to a trough. You may assume that the gradient wind approximation is valid.

9.4 (a) Use the wind observations at 500 hPa at 12 UTC 16 Feb 2003 at Blacksburg, Virginia and Wallops Island, Virginia to estimate the ageostrophic component of the flow due to the change in geostrophic wind speed between these two locations. You may assume that the reported winds at both radiosonde sites are equal to the geostrophic wind.

(b) Calculate the divergence in the left and right entrance regions of this jet streak by assuming that the ageostrophic flow at 500 km on either side of the jet is $0 \, \text{m} \, \text{s}^{-1}$.

(c) Are the regions of convergence and divergence calculated in part (b) consistent with the discussion in Section 9.2.1?

9.5 (a) Identify the location of the troughs, ridges, and jet streaks at the 300 hPa level for 00 UTC on (i) 14 February 2003 and (ii) 16 February 2003. Use the ETA model analyses on the CD-ROM.

(b) Is the change in sea level pressure between 00 UTC and 12 UTC 14 Feb 2003 in south-eastern Colorado consistent with the divergence of the ageostrophic winds due to the troughs, ridges, and jet streaks identified in part (a)? If not, what additional factors could cause this surface low-pressure system to change as observed?

(c) Is the change in sea level pressure between 00 UTC and 12 UTC 16 Feb 2003 in Mississippi and Alabama consistent with the divergence of the ageostrophic winds due to the troughs, ridges, and jet streaks identified in part (a)? If not, what additional factors could cause this surface low-pressure system to change as observed?

9.6 In a deepening mid-latitude cyclone, a horizontal divergence of $-3.3 \times 10^{-5} \, \text{s}^{-1}$ is observed at the surface. Assuming that this divergence is constant with height in the lowest 1000 m of the atmosphere, calculate the vertical velocity at a height of 1000 m. You may assume that the vertical velocity is $0 \, \text{m} \, \text{s}^{-1}$ at the surface.

9.7 Calculate the vertical velocity in the right and left entrance regions of the jet streak described in question 9.4. Assume that the tropopause is located 1 km above the jet streak and that there is no vertical motion at the tropopause.

9.8 Consider the surface low-pressure system at 00 UTC 16 Feb 2003 (see the weather maps on the CD-ROM). Use the thermal steering principle to determine the likely direction of travel of the cyclone.

Part II Atmospheric phenomena

10 Boundary layers

In Part I, we explored states of balance and adjustments to those states in what we call the *free* atmosphere. By this we mean that we have implicitly or explicitly excluded the effects of the viscous forces introduced in Section 4.3.3. However, it was problems with the assumption of no viscosity that slowed the acceptance of the mathematical theories of fluid flow amongst the experimenters and engineers who worked with real fluids and saw first-hand the effects of friction. This prompted Nobel Laureate Sir Cyril Hinshelwood, the British chemist, to observe at the start of the twentieth century that 'fluid dynamicists are divided into hydraulic engineers who observe things that cannot be explained, and mathematicians who explain things that cannot be observed'.

A serious problem in the study of these *inviscid* flows is that the theory predicts that in a flow without vorticity there can be no *drag* (that is, a pressure differential between one side of a moving body and the other). However, engineers attempting to build flying machines in the nineteenth century knew that real fluids such as air do produce drag. Otherwise, for example, parachutes would be useless. This conclusion was known as *d'Alembert's paradox*, after Jean Le Rond d'Alembert, the French mathematician who demonstrated this paradoxical aspect of the theory in 1768. D'Alembert himself saw that the conclusion was at odds with observation, writing 'I do not see then, I admit, how one can explain the resistance of fluids by the theory in a satisfactory manner.' The problem at the time was that without computers, there was no way to solve the Navier–Stokes equations without simplifying them as we did in Part I of the book. A crucial element of this simplification is the omission of the friction term. D'Alembert's paradox was finally resolved in 1904 by Ludwig Prandtl who found a way to connect the purely empirical engineering tradition and the purely theoretical mathematical tradition with this hypothesis:

> *Under a broad range of conditions, viscosity effects are significant, and comparable in magnitude with advection and other inertial forces, in layers adjoining solid boundaries and interfaces between fluids of different composition, and are small outside these layers.*

This hypothesis suggests that the effects of friction can be considered to be confined to a layer adjoining the boundary between atmosphere and ocean, or atmosphere and land. For researchers at the beginning of the twentieth century, this hypothesis meant that they could continue using simplifications that omitted friction, so long as there were no boundaries in their problem. This led to the concept of the free atmosphere, which we may now define formally as that part of the atmosphere above this *boundary layer*. This is why the approaches described in Part I led to some quite realistic models of the behavior of the real atmosphere.

It is important to note that this is an empirical hypothesis – that is, it does not arise from theoretical consideration of the equations of motion. However, the hypothesis has proved to be a useful perspective and yields credible results.

The existence of the boundary layer depends on the force of viscosity, but viscosity, a molecular process, acts on an extremely short spatial scale. In general this force, an expression of molecular diffusion, is several orders of magnitude smaller than the other terms in the momentum equation. We saw this in the scale analysis in Section 5.4.2 and quite reasonably concluded that we could omit this term from the equations of motion. However, this force does become comparable to the other terms in a very shallow layer directly adjoining the boundary – this is known as the *viscous sublayer*, and in the atmosphere it is generally on the order of millimeters deep.

However, the boundary layer hypothesized by Prandtl is not just a few millimeters deep – in the atmosphere it can be as deep as 1 kilometer. In fact, the role of viscosity is an indirect one: it causes a *no-slip boundary condition* (that is, the wind is moving at the same velocity as the boundary) at the surface or interface. This causes a large wind shear, which is one method of generating *turbulent eddies*. The eddies transfer heat and momentum much more efficiently than molecular diffusion. It is turbulence that defines the extent of Prandtl's boundary layer, and it is turbulence we must include in our equations of motion for the boundary layer.

10.1 Turbulence

Turbulence is difficult to describe, but easy to recognize. Properties on which most scientists agree are that it is three-dimensional and rotational in form, dissipative (that is, it converts kinetic energy to heat energy), and nonlinear. In practical terms, as we will see, the action of turbulence cannot be described by a differential equation.

In contrast with the motions we have studied so far, turbulent eddies have horizontal and vertical length scales of the same order of magnitude. These length scales can range from 10^{-3}m to 10^3m. In the boundary layer, all of these scales are of critical importance at all times. Hence, simplification through scale analysis is not possible. Because of this, the observational study of and theoretical explanations for turbulence are challenging. British meteorologist Lewis Fry Richardson suggested in 1920 that turbulence is a process where the main energy containing eddies receive kinetic energy directly, and then pass it down in a 'cascade' of eddy sizes to small eddies which dissipate it into heat. He summarized this process in a poem:

Big whorls have little whorls,
that feed on their velocity,

> *And little whorls have lesser whorls,*
> *And so on to viscosity.*

This cascade is caused by the stretching and twisting of the eddies. Because of it, if the motion is not to fade away there must be continuous production of turbulence to balance the dissipation. Making progress in simplifying this picture, Soviet scientist Andrei Nikolaevich Kolmogorov showed in 1941 that the only parameter needed to describe a turbulent flow is the average energy dissipation. A necessary condition for this was to consider only length scales far from both the scale at which energy is pumped into the flow and the scale at which the energy dissipates from the flow as heat. Importantly, he showed that it did not matter where you were in the fluid or in what direction you were traveling (a characteristic termed *isotropy*).

It turns out that isotropy is not always precisely maintained (see Section 10.3). However, it is approximated enough of the time that his important insight allows us to simplify our approach tremendously. This is because it makes clear that we do not need to worry too much about the precise details of the turbulent flow, just its overall effect on heat, momentum, and energy.

Despite the progress that has been made since Prandtl's revolutionary hypothesis, the problem of turbulence remains an extraordinarily difficult one. This was expressed in 1932 by the British physicist Horace Lamb, who is reported to have said,

> *I am an old man now, and when I die and go to Heaven there are two matters on which I hope for enlightenment. One is quantum electrodynamics, and the other is the turbulent motion of fluids. And about the former I am really rather optimistic.*

A very similar quote, however, is attributed to German physicist Werner Heisenberg, and so this may be apocryphal. Nevertheless, it reflects the fundamental challenge that turbulence represents, even today. In this chapter, we will consider some basic conclusions arising from the study of turbulent boundary layers.

10.2 Reynolds decomposition

We have seen that it is not feasible to study the precise flow details in turbulent motion, and in fact that it is not necessary. Hence, the complexity can be dealt with through a process of averaging known as *Reynolds decomposition*. This procedure was developed by British scientist Osborne Reynolds in the 1890s and has remained a powerful tool in the study of turbulent flows. In Reynolds decomposition, we represent any variable in a flow field (such as the wind velocity or temperature) by a slowly varying 'basic' component and a rapidly varying turbulent component. For example,

$$w = \overline{w} + w'$$

where $\overline{w}$ is the basic wind speed in the z direction and w' is the turbulent component. To properly represent the basic state, we must average the flow over a period of time which is long enough to average out all of the turbulent fluctuations but short enough to preserve any long-term trends and cycles. Generally, turbulent fluctuations occur over seconds and minutes, while synoptic development occurs on a time scale of hours. This suggests an averaging time somewhere between, such as 30 minutes. By definition then, the time average of the turbulent terms is zero. So in the case of the product of two variables, for example,

$$\overline{wT} = \overline{(\overline{w}+w')(\overline{T}+T')} = \overline{\overline{w}\,\overline{T}} + \overline{\overline{w}T'} + \overline{w'\overline{T}} + \overline{w'T'}$$
$$= \overline{w}\,\overline{T} + \underbrace{\overline{w}\,\overline{T'}}_{0} + \underbrace{\overline{w'}\,\overline{T}}_{0} + \overline{w'T'}$$
$$= \overline{w}\,\overline{T} + \overline{w'T'}$$

Note that the time average of the *product* of two turbulent terms (the second term above) is *not* necessarily zero. This term is most commonly called the *covariance term* and we can interpret this mathematically or physically. In the mathematical sense, if on average the turbulent vertical velocity is upward where the temperature is larger than average and downward where it is smaller, the product is positive and the variables are said to be positively correlated. From a more physical perspective, where $\overline{w'T'}$ is positive, vertical turbulent motions are transporting relatively warmer air upward and relatively cooler air downward – these both result in a net flux of warmer air upward. This net transport also applies to any water vapor that is present, so long as the air does not reach saturation. Because these covariance terms are a direct expression of the turbulent component of the flow, they are by Prandtl's hypothesis small in the free atmosphere, but significant in the boundary layer.

To apply Reynolds decomposition to the Navier–Stokes equations, first consider the total derivative in the x direction:

$$\frac{Du}{Dt} = \frac{\partial u}{\partial t} + u\frac{\partial u}{\partial x} + v\frac{\partial u}{\partial y} + w\frac{\partial u}{\partial z}$$

If we make an assumption, for simplicity, that the flow is non-divergent (and as we have seen, this is a rather strict assumption), the continuity equation is

$$\frac{\partial u}{\partial x} + \frac{\partial v}{\partial y} + \frac{\partial w}{\partial z} = 0$$

Since this expression is zero, we can add it, multiplied by u, to the total derivative with no impact:

$$\frac{Du}{Dt} = \frac{\partial u}{\partial t} + u\frac{\partial u}{\partial x} + v\frac{\partial u}{\partial y} + w\frac{\partial u}{\partial z} + u\left(\frac{\partial u}{\partial x} + \frac{\partial v}{\partial y} + \frac{\partial w}{\partial z}\right)$$
$$= \frac{\partial u}{\partial t} + \frac{\partial u^2}{\partial x} + \frac{\partial uv}{\partial y} + \frac{\partial uw}{\partial z}$$

In order to Reynolds-decompose this expression, we separate each term into the mean and fluctuating parts, and take the average of all terms (the details of this are shown on the CD-ROM). The Reynolds-decomposed expression is then

$$\overline{\frac{Du}{Dt}} = \frac{D\overline{u}}{Dt} + \frac{\partial}{\partial x}\left(\overline{u'u'}\right) + \frac{\partial}{\partial y}\left(\overline{u'v'}\right) + \frac{\partial}{\partial z}\left(\overline{u'w'}\right)$$

where $\overline{D}/Dt$ is the rate of change following the mean motion. Using this term in the zonal momentum equation, where for simplicity we denote the friction term, representing molecular diffusion, from Equation (4.6) by F_{rx},

$$\overline{\frac{Du}{Dt}} = -\overline{\frac{1}{\rho_{00}}\frac{\partial p}{\partial x}} + \overline{fv} + \overline{F_{rx}}$$

$$\frac{D\overline{u}}{Dt} + \frac{\partial}{\partial x}\left(\overline{u'u'}\right) + \frac{\partial}{\partial y}\left(\overline{u'v'}\right) + \frac{\partial}{\partial z}\left(\overline{u'w'}\right) = -\frac{1}{\rho_{00}}\frac{\partial}{\partial x}(\overline{p}+p') + f(\overline{v}+v') + (\overline{F}_{rx}+F'_{rx})$$

$$\frac{D\overline{u}}{Dt} = -\frac{1}{\rho_{00}}\frac{\partial \overline{p}}{\partial x} + f\overline{v} + \overline{F}_{rx} - \left[\frac{\partial}{\partial x}\left(\overline{u'u'}\right) + \frac{\partial}{\partial y}\left(\overline{u'v'}\right) + \frac{\partial}{\partial z}\left(\overline{u'w'}\right)\right] \quad (10.1)$$

Thus, it is clear that the friction term is distinct from the turbulence term. This term, because it is the sum of the gradients (that is, the divergence) of a product (that is, a flux, Section 3.7), is usually known as the *turbulent flux divergence*. In the boundary layer, the turbulent flux divergence is on the same order of magnitude as the other terms in this equation. Hence, it cannot be neglected even when *only* the basic flow is of direct interest.

Using the same process, we can write down the meridional momentum equation:

$$\frac{D\overline{v}}{Dt} = -\frac{1}{\rho_{00}}\frac{\partial \overline{p}}{\partial y} - f\overline{u} + \overline{F}_{ry} - \left[\frac{\partial}{\partial x}\left(\overline{v'u'}\right) + \frac{\partial}{\partial y}\left(\overline{v'v'}\right) + \frac{\partial}{\partial z}\left(\overline{v'w'}\right)\right] \quad (10.2)$$

It is also useful, in the context of the boundary layer, to include some kind of prediction for the temperature structure. For simplicity, we will assume that all motion is adiabatic, and hence potential temperature is conserved. That is, the material derivative of potential temperature is zero. Assuming a potential temperature structure $\theta_0(z)$ associated with the hydrostatic profile, this can be written

$$\frac{D}{Dt}(\theta + \theta_0(z)) = 0$$

$$\frac{D\theta}{Dt} + w\frac{\partial \theta_0}{\partial z} = 0$$

which in Reynolds-decomposed form becomes

$$\frac{D\overline{\theta}}{Dt} = -\overline{w}\frac{\partial \theta_0}{\partial z} - \left[\frac{\partial}{\partial x}\left(\overline{u'\theta'}\right) + \frac{\partial}{\partial y}\left(\overline{v'\theta'}\right) + \frac{\partial}{\partial z}\left(\overline{w'\theta'}\right)\right] \quad (10.3)$$

Equations (10.1)–(10.3) are equations to calculate the wind and temperature in the boundary layer. However, with the tools we have available, we cannot solve these equations, even numerically. This is because we have introduced three additional independent variables for each equation. For example, in Equation (10.1), these variables are $\overline{(u'u')}$, $\overline{(u'v')}$, and $\overline{(u'w')}$. In Equation (10.2), we have only two additional variables, since $\overline{(u'v')} = \overline{(v'u')}$, but in Equation (10.3) we again have three new variables. Hence, to solve this system of equations we must develop *eight* additional equations which either do not include any new variables, or include additional equations for any introduced variables! Such a process is known as *closure*. Closure schemes are named by the highest order equations that are retained. For example, a first-order scheme is one in which Equation (10.1), the basic flow equation, is used and the turbulent fluxes are expressed only in terms of the basic state. This necessitates assumptions regarding the relationship between the basic state and the structure of the turbulent flow. These are known as *closure assumptions*, to be discussed in Section 10.4. A second-order closure derives additional prognostic equations for the turbulent flux terms, but approximates any higher order perturbations. And so forth.

10.3 Generation of turbulence

For a turbulent flow to be maintained, continuous production of turbulence is required to balance the dissipation. This means that, somehow, kinetic energy must be imparted to the largest eddies in the flow. This turbulence generation can take place in two ways. The first is through convection: when the temperature profile is unstable, air must start to circulate vertically to redistribute heat upward until a stable profile is achieved. This convection gives rise to large eddies that can then feed the turbulence through the energy cascade. Of course, if the temperature profile is very stable, this will act to suppress turbulence. One may think of this process as a conversion from basic flow potential energy, embodied by the potential temperature profile, to turbulent kinetic energy, through the action of buoyancy forces. Mathematically, this can be written

$$\overline{(w'\theta')} \frac{g}{\theta_0}$$

where w' is the turbulent vertical wind, θ' is the turbulent component of the potential temperature variation, g is the acceleration due to gravity, and θ_0 is the basic flow potential temperature profile. This term may be positive (for turbulence generation) or negative (when a stable profile suppresses turbulence). The second is through mechanical processes, due to the presence of wind shear. Because of the no-slip boundary condition imposed by the viscous force, large gradients of wind are present at the boundary. This is an unstable situation, and so large turbulent eddies are generated to decrease the wind shear. This can be understood as a conversion from basic flow kinetic energy to turbulent kinetic energy, and can be written

$$-\overline{(u'w')} \frac{\partial \overline{u}}{\partial z} - \overline{(v'w')} \frac{\partial \overline{v}}{\partial z}$$

where u, v, and w have their usual meaning as wind components. This term is always positive. Finally, like all aspects of fluid flow, turbulence can also be modified by transport – that is, it can be advected like temperature or basic flow momentum.

The source of turbulence in any particular situation will affect the structure of the turbulence. In general, turbulence caused through mechanical generation has smaller scales than turbulence caused by buoyant production. Thus, when both convective turbulence and mechanical turbulence are present, relatively slow and large-scale fluctuations are superimposed on relatively fast and small-scale variations, resulting in very complex patterns. The large eddies tend to collect the smaller turbulent circulations into their upward branches, with the result that these regions of the convective boundary layer contain more turbulent kinetic energy than the downward branches of the large eddies. Further, fluctuations in wind direction are amplified by convective turbulence. It is also known that, close to the ground, all turbulent eddies tend to be small. However, mechanically generated eddies near the ground are elongated in the direction of the wind, whereas convective eddies are more isotropic.

The relative importance of buoyant production (or loss through stability) and mechanical production in a particular turbulent flow field can be expressed as a mathematical ratio of the terms representing these processes. The ratio is usually written

$$-\left[\frac{\overline{(w'\theta')}\frac{g}{\theta_0}}{-\overline{(u'w')}\frac{\partial \bar{u}}{\partial z} - \overline{(v'w')}\frac{\partial \bar{v}}{\partial z}}\right]$$

This ratio is called the *flux Richardson number* R_f (after Lewis Fry Richardson of course). Experiments have shown that if this number is great than about $1/4$, the air is so highly stable that turbulence is suppressed even in the presence of wind shear. If R_f is less than 0, then the turbulence is maintained by convection. Between these two values, the mechanical production appears to be strong enough to sustain turbulent flow even in a stable profile.

10.4 Closure assumptions

Consider a simplified case in which we assume that the structure of the boundary layer is horizontally homogeneous (that is, $\partial/\partial x, \partial/\partial y = 0$). Equations (10.1)–(10.2) become

$$\frac{\overline{D\bar{u}}}{Dt} = -\frac{1}{\rho_{00}}\frac{\partial \bar{p}}{\partial x} + f\bar{v} - \frac{\partial}{\partial z}\overline{(u'w')} + F_{rx}$$

$$\frac{\overline{D\bar{v}}}{Dt} = -\frac{1}{\rho_{00}}\frac{\partial \bar{p}}{\partial y} - f\bar{u} - \frac{\partial}{\partial z}\overline{(v'w')} + F_{ry}$$

Synoptic scaling is still applicable for synoptic scale motions in the boundary layer, except that according to Prandtl's hypothesis the turbulent component must scale as large as the other terms in the equations. If we also assume we are outside the viscous sublayer, then synoptic scaling allows us to neglect the friction terms F_{rx} and F_{ry}. These scale considerations result in

$$0 = -\frac{1}{\rho_{00}}\frac{\partial \overline{p}}{\partial x} + f\overline{v} - \frac{\partial}{\partial z}\left(\overline{u'w'}\right)$$

$$0 = -\frac{1}{\rho_{00}}\frac{\partial \overline{p}}{\partial y} - f\overline{u} - \frac{\partial}{\partial z}\left(\overline{v'w'}\right)$$

and the equations can be expressed in terms of the geostrophic wind:

$$0 = f\left(\overline{v} - \overline{v}_g\right) - \frac{\partial}{\partial z}\left(\overline{u'w'}\right)$$

$$0 = -f\left(\overline{u} - \overline{u}_g\right) - \frac{\partial}{\partial z}\left(\overline{v'w'}\right) \qquad (10.4)$$

To close this set of equations, we need expressions to determine the two remaining turbulent flux terms. The assumptions made to devise these expressions determine the 'order' of the closure scheme. Hence, a scheme which assumes the turbulent flux terms are constant is called a 'zero-order' scheme. One that approximates the turbulent flux in terms only of the mean flow is called a 'first-order' scheme. If the turbulent flux terms are represented by their own additional equations, with higher order fluxes then approximated, the scheme becomes 'second-order', and so on. The higher the order of closure scheme, the more degrees of freedom and hence the potentially more accurate (if the assumptions made are appropriate) the scheme will be.

Consider the situation illustrated in Figure 10.1 in which a convective layer, heated from below on a warm spring afternoon, is topped by a highly stable layer. In such a situation, turbulent mixing can lead to the formation of a *well-mixed layer* in which the wind speed and potential temperature are nearly independent of height.

Observations indicate that in such a situation, we can represent the surface turbulent momentum flux by

$$\overline{(u'w')}_s = -C_d \left|\overline{\overline{u}}\right|\overline{u} \qquad \overline{(v'w')}_s = -C_d \left|\overline{\overline{u}}\right|\overline{v}$$

where $\left|\overline{\overline{u}}\right| = \sqrt{\overline{u}^2 + \overline{v}^2}$. That is, the turbulent flux is proportional to the square of the basic wind speed, and the constant of proportionality C_d is obtained empirically. We assume that the turbulent momentum flux at the entrainment layer is zero because this is the point at which, moving into the free atmosphere, there should no longer be significant energy in the turbulent component. This is known as a 'half-order' closure scheme since we have not only assumed the turbulent flux terms are a function of the basic flow, but also assumed the shape of the basic flow profile in advance. Hence, we have reduced the degrees of freedom from a basic first-order scheme, but not so greatly as to create a zero-order scheme.

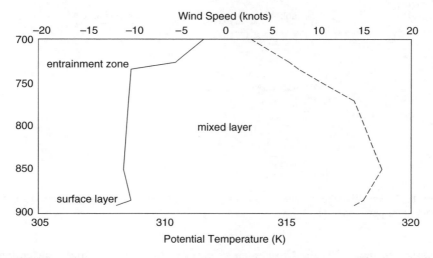

Figure 10.1 Atmospheric sounding for Amarillo, Texas at 6 p.m. local time on 26 May 2005. The profiles of wind (dashed line) and potential temperature (solid line) from the surface (at 1099 m) to 700 hPa are shown. The surface layer acts as the transition from the no-slip boundary condition to the boundary layer. The entrainment zone is the region where the boundary layer air mixes with the free atmosphere. Between lies the mixed layer, where the wind and temperature are approximately uniform

We can then integrate Equations (10.4) from the surface to the height h of the entrainment layer to find

$$\bar{u} = \bar{u}_g - \frac{C_d |\vec{u}| \bar{v}}{fh} \qquad \bar{v} = \bar{v}_g + \frac{C_d |\vec{u}| \bar{u}}{fh}$$

Thus, this model produces a cross-isobaric component to the wind (that is, an ageostrophic wind) in the well-mixed layer which does work to balance the frictional dissipation at the surface. This is known as the *mixed layer theory*.

Consider as an illustration a case in which the geostrophic flow is purely zonal and in the positive x direction (Figure 10.2). In this case, $\bar{v}_g = 0$, and the basic meridional wind $\bar{v}$ is a small component in the northward direction. Since the geostrophic wind is westerly, this meridional wind must be blowing toward low pressure. The zonal basic flow $\bar{u}$ is modified also, through a small decrease arising from the contribution of the cross-isobaric flow. Thus, this model produces the type of low-level slowing and turning toward low pressure that we have come to expect from our examination of weather charts.

Since generally the boundary layer turbulence tends to reduce the wind speed, the turbulent momentum flux terms are often called boundary layer friction or turbulent drag, but this 'friction' must not be confused with processes involving molecular viscosity.

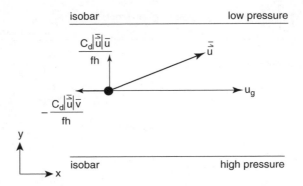

Figure 10.2 Resulting cross-isobaric flow in a well-mixed layer in the Northern Hemisphere ($f > 0$) in the case of pure zonal flow

Example In June 1893 the Norwegian explorer and scientist Fridtjof Nansen loaded 6 years' supply of food and 8 years' supply of fuel onto the ice-strengthened ship *Fram* and left Oslo port, heading east to Siberia. His plan was to sail northward from Siberia, and let his ship be frozen into the ice when the Arctic Sea froze, and drift with the ice, letting it carry him to the North Pole and then to Greenland. This plan was based on his observations that surface ocean currents were not in the same direction as the prevailing winds, but between 20° and 40° to the right of the wind's direction. Although he did not make it all the way to the North Pole, his expectations proved correct, and three years later the *Fram* reached Spitsbergen, finally leaving the ice and reaching Norway in August 1896 (see the map of his voyage in the CD-ROM).

How can boundary layer theory explain the discrepancy between the winds and the surface currents that Nansen observed?

After hearing of Nansen's observations, V. Bjerknes invited his student V. Walfrid Ekman to use the new boundary layer theory to understand the discrepancy. In doing so, Ekman had to account for the possibility of the presence of strong wind shear, which occurs frequently in the Arctic. Hence, rather than using mixed layer theory, he assumed that a turbulent flow transports heat and momentum in a manner analogous to molecular diffusion:

$$\overline{u'w'} = -K_m \frac{\partial \overline{u}}{\partial z}$$

$$\overline{v'w'} = -K_m \frac{\partial \overline{v}}{\partial z} \qquad (10.5)$$

where K_m is the *eddy viscosity coefficient*, analogous to ν, the molecular viscosity coefficient (Section 4.3.3). Since turbulence is more effective at mixing than viscosity, one would expect that in such a formulation $K_m \gg \nu$. This is a first-order closure scheme since the flux terms are approximated using only the basic flow quantities, but no assumptions are made about the form of the basic flow. In actuality, we have made

CLOSURE ASSUMPTIONS

some assumptions, since of course we have assumed synoptically scaling, horizontally homogeneous flow. However, this is typical enough that it is not considered to reduce the order of the closure.

Ekman further assumed no significant pressure gradients, which is better approximated in the ocean than in the atmosphere. Assuming some as yet unknown formulation for the eddy viscosity coefficients, Equations (10.4) become

$$f\bar{v} + K_m \frac{\partial^2 \bar{u}}{\partial z^2} = 0$$

$$-f\bar{u} + K_m \frac{\partial^2 \bar{v}}{\partial z^2} = 0 \qquad (10.6)$$

Equations (10.6) have come to be known as the *Ekman boundary layer equations*. These equations can be solved to determine the currents in the ocean being driven by a surface wind.

To find the solutions to these equations, we must define some boundary conditions and postulate a form of the solution. We start by aligning our horizontal axes with the surface current so that $\vec{\bar{u}}(z=0) = (\bar{u}_0, 0)$. The other boundary condition is

$$\bar{u} \to 0 \text{ and } \bar{v} \to 0 \text{ as } z \to -\infty$$

We can postulate a form for the solution that conforms with this boundary condition. To do so, we define a constant (and hence make the implicit assumption that K_m is constant) which will make our equations simpler to read:

$$h = \sqrt{\frac{2K_m}{f}} \qquad (10.7)$$

and then write

$$\bar{u}(z) = A e^{\frac{z}{h}} \cos\left(\frac{z}{h}\right)$$

$$\bar{v}(z) = B e^{\frac{z}{h}} \sin\left(\frac{z}{h}\right)$$

The surface ($z=0$) boundary condition gives us $A = \bar{u}_0$ but does not allow us to define B. We do this by substituting the assumed solution back into one component of Equations (10.6). We choose the second component, but the first would give the same result:

$$\bar{u} = \frac{h^2}{2} \frac{\partial^2 \bar{v}}{\partial z^2}$$

$$\bar{u}_0 e^{\frac{z}{h}} \cos\left(\frac{z}{h}\right) = \frac{h^2}{2} \frac{2B}{h^2} e^{\frac{z}{h}} \cos\left(\frac{z}{h}\right)$$

$$\Rightarrow B = \bar{u}_0$$

Hence, we have derived the current profile

$$\bar{u}(z) = \bar{u}_0 e^{\frac{z}{h}} \cos\left(\frac{z}{h}\right)$$
$$\bar{v}(z) = \bar{u}_0 e^{\frac{z}{h}} \sin\left(\frac{z}{h}\right) \qquad (10.8)$$

This profile, of course, does not tell us the relationship between the current at the surface and the wind speed above it. To determine this, we must apply a further boundary condition at the surface. Consider the following.

How does a fluid deform? By definition, the distinguishing property of a fluid is that it cannot support a shearing stress (Section 3.1). Thus, at the surface, the stress must be continuous across the air–water interface. Hence, the stress in the air, the wind stress, must be equal to the stress in the water. This replaces the 'no-slip' boundary condition introduced at the start of this chapter that is appropriate when the boundary is a solid surface. So, if we calculate the water stress, this will give us the wind stress, and hence the direction of the wind.

We have assumed a form already for the stress in the water in Equations (10.5). To write these in terms of stress directly, at the surface we have

$$\tau_x = \rho K_m \left.\frac{\partial \bar{u}}{\partial z}\right|_0 = \frac{\rho K_m \bar{u}_0}{h}$$
$$\tau_y = \rho K_m \left.\frac{\partial \bar{v}}{\partial z}\right|_0 = \frac{\rho K_m \bar{u}_0}{h} \qquad (10.9)$$

Hence, assuming the surface current is purely zonal and toward the east, the surface wind must be directed toward the north-east, or 45° to the left of the current.

This explains what Nansen observed on his ship.

At a greater depth, although the speed of the current will decrease, the direction of the current will deviate even further from the wind's direction. Eventually, at a certain depth, the sea water may even flow upwind. The form of the current profile is called the *Ekman spiral* (Figure 10.3).

While American scientist Kenneth Hunkins was probably the first researcher to observe an Ekman spiral in the real ocean, from *Ice Station Alpha* during the International Geophysical Year 1957–1958 (see Figure 10.3), routine measurements capable of discerning both current speed and direction from a stable platform were not possible until the 1980s. Observed spirals tend to turn less with depth than the Ekman model predicts, but nevertheless the agreement between this simple theory and the observations is substantial.

The height of the atmospheric mixed layer, or depth of the oceanic mixed layer, is easily defined by the extent of uniform properties, particularly potential temperature, close to the boundary. In the case of the Ekman spiral, a depth of the layer can also be defined. For example, at the depth

$$z = -\pi h$$

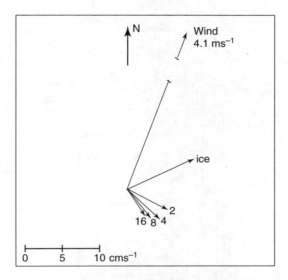

Figure 10.3 A wind-driven current velocity spiral as measured using an instrument, lowered from an ice floe, that measures current as averages over around 15 minutes. The numbers show the depth (m). From Hunkins (1966), perhaps the first field observation of an Ekman spiral in the ocean. Reprinted from *Deep Sea Research*, **13**, K. Hunkins, 'Ekman drift currents in the Arctic Ocean', 607–620, © 1966, with permission from Elsevier

the currents are

$$\bar{u} = -0.04\bar{u}_0 \text{ and } \bar{v} = 0$$

Similarly,

$$\bar{u}|_{-\frac{3\pi h}{2}} = 0 \text{ and } \bar{v}|_{-\frac{3\pi h}{2}} = -0.008\bar{u}_0$$

$$\bar{u}|_{-2\pi h} = 0.001\bar{u}_0 \text{ and } \bar{v}|_{-2\pi h} = 0$$

and so on, as the solution oscillates closer and closer to zero. Typically, the value $z = -\pi h$ is taken to be a measure of the depth of the boundary layer. As expected by the force balance between wind stress and Coriolis force, the depth depends on the latitude and the eddy viscosity coefficient.

Turning again to the theoretical form of the spiral (Figure 10.4), it is apparent that while the direction of the current changes with depth, integrated over the entire water column the amount of water flowing upwind and downwind cancel, so that on average there is no current either upwind or downwind. The average direction of the current is at 90° to the right of the wind's direction. This net current is called the *Ekman transport*.

This net current has important implications for the observed oceanic circulation. Because, at the middle latitudes, the oceans are often dominated by large, persistent,

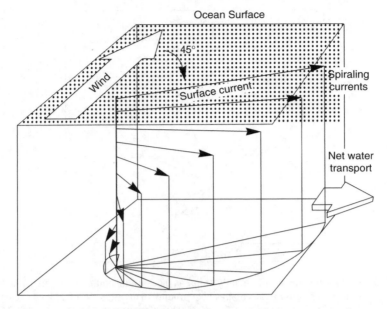

Figure 10.4 Schematic showing the Ekman wind-driven current velocity spiral

high-pressure systems driving an anticyclonic flow, the Ekman transport directed to the right of the wind stress predicts a transport of water into the center of the vortex, or gyre. This convergence at the surface causes elevated sea level heights at the center of the gyre, as is actually observed. The surface convergence also drives downward motion and divergence deep below the center of the gyre, creating a circulation. This secondary circulation is often referred to as Ekman pumping.

10.4.1 The analogy between molecular and turbulent diffusion

For the Ekman solution obtained above, an important assumption was made: that the form of turbulent momentum transfer is analogous to molecular diffusion as described in classical kinetic theory. The earliest and most popular form of this analogy was the 'mixing length theory', not to be confused with the 'mixed layer theory' described above. The simplest form of this theory assumed that turbulent transfer was caused by the motion of discrete parcels of fluid, like molecules, moving relative to one another. Hence, a parcel originates from one location with a set of properties (for example, momentum, temperature) and moves a certain distance at constant velocity before mixing with its surroundings. From the resulting mixture, a new parcel is identified which repeats the process anew. This distance is called the *mixing length*, an analog of the 'mean free path' in kinetic theory, and in its simplest incarnations yields a constant eddy diffusivity coefficient K_m.

Fundamental differences between the diffusion of momentum by molecular viscosity and by turbulence were identified as early as 1895 by Osborne Reynolds. Molecular velocities are very large, on the same order of magnitude as the speed of sound, but molecular diffusion is very small because collisions act constantly to restore thermodynamic equilibrium, and it is only over the very small mean free path before the collision that the molecule carries identifiable properties. In contrast, fluid parcel velocities are relatively slow, but kinetic energy is extracted from the mean flow and transferred to the turbulent flow rather rapidly. This is reflected in empirically determined values of K_m, which are flow dependent but usually of the order of $10^1 \text{m}^2\text{s}^{-1}$, in sharp contrast to values of ν, which are generally of the order of 10^{-5} to $410^{-6} \text{m}^2\text{s}^{-1}$.

Such contrasts demonstrate the important scale separation between the statistics of molecular velocities in kinetic theory and the statistics of fluid velocities in turbulence. They also illustrate why the concept of a mean free path in macroscopic flows is unlikely. Nevertheless, refinements to this approach have continued over the years, particularly with regard to approaches known as *similarity theory*, which involve an empirical determination of dimensionless constants or functions. More recent approaches have returned to the statistical description of turbulent fluxes, sometimes combined with the explicit treatment of very large convective eddies that can have length scales as large as the depth of the boundary layer.

Review questions

10.1 The following observations of w and θ were made at 1 second intervals during a research flight over Kansas during a summer afternoon:

$w = 1.1, 1.5, 1.4, 0.9, 1.0, 1.2, 0.8, 0.7, 1.3,$ and 1.1 m s^{-1}

$\theta = 290.5, 291.0, 290.8, 290.3, 290.5, 291.0, 290.3, 289.9, 291.1,$ and 290.7 K

(a) Using Reynolds decomposition calculate the basic and turbulent component for each time series.

(b) Calculate the covariance term, $\overline{w'\theta'}$, based on the time series.

(c) Discuss the physical implications of the sign of the covariance term calculated in part (b).

10.2 Following the method used to derive Equation (10.1) show the derivations for Equations (10.2) and (10.3).

10.3 Discuss how Figure 10.2 would change for

(a) $C_D = 0$

(b) C_D smaller than used in Figure 10.2

(c) C_D larger than used in Figure 10.2.

10.4 Plot the Ekman profiles of $\bar{u}, \bar{v}$, and $\left|\vec{\bar{u}}\right|$ in the ocean, from the surface to a depth of $-2\pi h$, for a surface current of $10\,\text{cm}\,\text{s}^{-1}$ and $h = 100\,\text{m}$.

10.5 Solve the Ekman equations for the basic flow in the atmospheric boundary layer. Choose as your boundary conditions a 'no-slip condition' at the surface, and

$$\bar{u} \to \bar{u}_g \text{ and } \bar{v} \to \bar{v}_g \text{ as } z \to \infty$$

11 Clouds and severe weather

Water exists in the atmosphere in three forms – as water vapor (a gas), as liquid water, and as ice. Whereas water vapor is transparent to visible light and hence cannot be seen in the air, liquid water and ice are readily seen in the atmosphere, as clouds, rain, hail, and snow. It is the presence of water in the atmosphere that gives our weather its character. Many of the most severe weather events, such as hurricanes and thunderstorms, and also many of the beautiful atmospheric phenomena, are associated with water. Rainbows are of course associated with refraction through raindrops. Ice crystals in the clear air can also produce interesting effects such as sun dogs, halos, and pillars. In this chapter, the role of moist processes in generating some examples of severe weather will be explored.

11.1 Moist processes in the atmosphere

In Section 3.3.2, parameters were introduced that describe the amount of water vapor in the air, including the water vapor pressure e and the mixing ratio r. These concepts will be expanded here to help describe the formation of clouds in the atmosphere.

If the masses of water vapor and dry air in the mixture are known, the vapor pressure can be calculated using Dalton's law (Section 3.3.1), by

$$e = \frac{\text{moles of water vapor}}{\text{moles of water vapor and dry air}} p$$

$$= \frac{m_v/M_v}{m_v/M_v + m_d/M_d} p$$

Since the mixing ratio is simply $r = m_v/m_d$, this can be written

$$e = \frac{r}{r + M_v/M_d} p \tag{11.1}$$

The ratio M_v/M_d is a constant, usually denoted ε, having a value of 0.622.

Applied Atmospheric Dynamics   Amanda H. Lynch, John J. Cassano
© 2006 John Wiley & Sons, Ltd

Equation (11.1) can be rearranged, so that

$$r = \frac{\varepsilon e}{p - e} \quad (11.2)$$

can be used to calculate the mixing ratio from a known value of vapor pressure.

Clouds form in the atmosphere when the water vapor pressure exceeds the *saturation vapor pressure*, e_s, at that temperature. *Saturation* is, in fact, a dynamic equilibrium in which there are as many molecules returning to the liquid water (condensing) as there are escaping (evaporating). At this point the vapor is said to be saturated, and the vapor pressure is called the saturated vapor pressure. Equation (11.2) can be used with e_s to calculate the saturation mixing ratio, r_s. Since the molecular kinetic energy is greater at higher temperature, more molecules can escape the surface and the saturated vapor pressure and saturation mixing ratio are correspondingly higher.

There are a large number of semi-empirical equations that have been developed to calculate the saturation vapor pressure as a function of temperature. The Goff–Gratch equation (see Figure 11.1) is generally considered to be the reference, but many other equations are in use in the scientific community. The differences between these formulations at typical tropospheric temperatures and pressures are too small to concern us here, although they can be important in the upper troposphere and above. One of the simpler semi-empirical equations, which is used to calculate the saturation vapor pressure over a liquid water surface, is known as *Teten's formula*:

$$e_s = e_0 \exp\left(\frac{b(T - T_1)}{T - T_2}\right) \quad (11.3a)$$

where $e_0 = 6.112\,\text{hPa}$, $b = 17.27\,\text{K}^{-1}$, $T_1 = 273.16\,\text{K}$, $T_2 = 35.86\,\text{K}$, and T is the air temperature, also with units of K. Similarly the vapor pressure, e, can be calculated with

$$e = e_0 \exp\left(\frac{b(T_d - T_1)}{T_d - T_2}\right) \quad (11.3b)$$

where we have replaced T with the dew point temperature T_d (Section 1.2.3). Equations (11.3a) and (11.3b) indicate that when $T = T_d$, $e_s = e$ and the air is saturated, as expected from the discussion of dew point temperature in Section 1.2.3.

When the air is in dynamic equilibrium with a surface of ice, rather than liquid water, a lower saturation vapor pressure results (Figure 11.1). This is because molecules escape from the surface of ice less readily than from liquid water. Hence, it is important to always specify whether the saturation vapor pressure is with respect to water or to ice. The ratio of the saturation vapor pressure over a liquid water surface to the saturation vapor pressure oven an ice surface, e_{si}, can be calculated using

$$\frac{e_s}{e_{si}} = \left(\frac{273\,\text{K}}{T}\right)^{2.66} \quad (11.4)$$

where T is given with units of K.

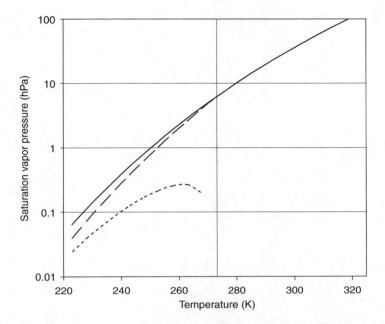

Figure 11.1 The saturation vapor pressure with respect to both water (solid line) and ice (dashed line), and the difference between them (dotted line), calculated using the Goff–Gratch equation. The vertical line indicates the freezing temperature

Finally, relative humidity is one measure of the amount of moisture in the air that is frequently reported in weather forecasts and discussions. This quantity is defined as

$$\text{RH} = \frac{e}{e_s} \times 100\% \approx \frac{r}{r_s} \times 100\% \tag{11.5}$$

When RH $= 100\%$, $e = e_s$ and the air is saturated with respect to a liquid water surface. We can also determine the relative humidity with respect to an ice surface, which is given by

$$\text{RH}_i = \frac{e}{e_{si}} \times 100\% \tag{11.6}$$

Example Use a skew T–$\log P$ diagram to determine the mixing ratio, saturation mixing ratio, and relative humidity at the surface at Fort Worth, Texas at 00 UTC 5 Jun 2005.

Figure 11.2 shows the skew T–$\log P$ diagram for Fort Worth on this date, which was a day that saw considerable thunderstorm activity across the central United States (see the projects section of the **CD-ROM**.) Skew T–$\log P$ diagrams are commonly used by meteorologists to diagnose the structure of the atmosphere at a particular location and to determine how air parcels will be modified due to ascent or descent.

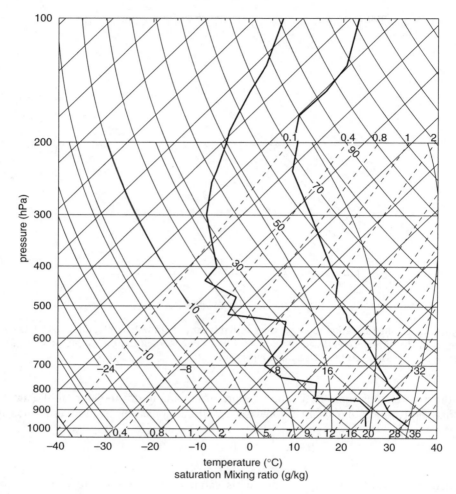

Figure 11.2 'Skew T–log P' diagram for a sounding taken at Fort Worth, Texas on 5 June 2005 at 00 UTC, showing the temperature and dew point temperature profiles

The logarithm of pressure is used as the vertical axis of the plot, and temperature is plotted using an axis that is skewed at a 45° angle to the horizontal (hence the name of the diagram). The saturation mixing ratio is also accorded a skewed axis, as are lines that represent dry and moist adiabatic ascent. The lines that represent dry adiabatic ascent, known as dry adiabats, are also lines of constant potential temperature (Section 3.4). The temperature and dew point temperature profiles as measured by a radiosonde (Section 1.3.1) are plotted on the diagram. The figure is also shown on the CD-ROM with color coding to help identify the axes.

The saturation mixing ratio can be determined by locating the saturation mixing ratio line that intersects a point defined by the measured temperature and pressure.

For this example, we are interested in the saturation mixing ratio at the surface. At the surface, the temperature at this time was 31 °C and the pressure was approximately 990 hPa. The point given by this temperature and pressure does not lie on any of the saturation mixing ratio lines plotted on the diagram, and so we will need to estimate the saturation mixing ratio based on the saturation mixing ratio lines that lie on either side of our point of interest. Based on Figure 11.2 we estimate the saturation mixing ratio for this point to be around 30 g kg^{-1}.

We will use the point defined by the surface dew point temperature (22 °C) and surface pressure (990 hPa) to determine the mixing ratio. To do this we first need to note that the saturation mixing ratio lines on a skew T diagram also represent lines of constant mixing ratio. The mixing ratio of the air parcel is given by the mixing ratio line that intersects the point defined by the dew point temperature and pressure of an air parcel. The point defined by the surface dew point temperature and pressure does not lie on any of the mixing ratio lines and so, as before, we need to estimate the mixing ratio based on the value of the mixing ratio lines that lie on either side of our point of interest. Doing this, we estimate the mixing ratio at the surface to be approximately 17 g kg^{-1}.

From the values of mixing ratio and saturation mixing ratio that we have estimated from the skew T–log P diagram we can calculate the relative humidity of the air at the surface using Equation (11.5):

$$RH \approx \frac{r}{r_s} \times 100\%$$

$$\approx \frac{17}{30} \times 100\%$$

$$\approx 57\%$$

There are two ways in which a parcel of air may become saturated. One is to add more water vapor to the air. More commonly, though, air becomes saturated because its temperature decreases. This may occur through *in situ* processes such as radiative cooling, or through ascent, as a parcel rises and cools adiabatically (Section 3.4). This ascent may be due to a flow over a mountain, turbulent mixing arising from convection, or low-level convergence in the vicinity of a front, for example. The level at which saturation occurs, in this case, is called the *lifting condensation level* (LCL). The LCL is a good estimate of the cloud base height for clouds that form through this process.

If ascent and cooling continue, condensation of the excess water will start to occur. Because of the heat released through this process, the cooling associated with this ascent is less than that observed with *dry adiabatic ascent*, and the potential temperature is no longer conserved. In observations, the liquid water content of a cloud is generally less than expected from the process described here, because water is lost from the cloud through mixing with surrounding unsaturated air, and through rainfall.

Example Find the lifting condensation level at Fort Worth, Texas at 00 UTC 5 Jun 2005.

To determine the lifting condensation level we will again use the skew T–$\log P$ diagram shown in Figure 11.2, and will consider an air parcel that is lifted from the surface. We will need to determine how the temperature and dew point temperature of this air parcel change as the parcel is lifted through the atmosphere, and identify the location at which the temperature and dew point temperature of the air parcel first become equal.

Since the air parcel is not initially saturated ($T \neq T_d$) we know that this air parcel will undergo dry adiabatic ascent (Section 3.4) when lifted, and will have a constant potential temperature. To represent the change in air temperature of the air parcel for this dry adiabatic process on the skew T–$\log P$ diagram we start from the surface atmospheric temperature and draw a line from the surface temperature parallel to the dry adiabats (see the color version of Figure 11.2 on the **CD-ROM**). Physically this represents a conservation of potential temperature during the dry adiabatic process, and each point on this line defines the temperature and pressure of the air parcel required to maintain a constant potential temperature during ascent.

We also need to consider how the dew point temperature of this air parcel will change during ascent. First, we note that during dry adiabatic ascent the total amount of water vapor in the air parcel remains constant, and thus the mixing ratio will also remain constant. In the previous example we found that the mixing ratio of the air parcel is given by the intersection of the mixing ratio lines with the parcel dew point temperature and pressure, and was equal to $17\,\text{g}\,\text{kg}^{-1}$ at the surface. To represent the conservation of mixing ratio we draw a line that starts at the value of the surface mixing ratio and is parallel to the mixing ratio lines. Each point along this line has a constant mixing ratio of $17\,\text{g}\,\text{kg}^{-1}$ and defines the dew point temperature and pressure of the air parcel as it rises through the atmosphere.

Where the two lines we have drawn intersect, the air parcel temperature and dew point temperature are equal. It is at this point that the rising air parcel first becomes saturated and a cloud will form, at the lifting condensation level (LCL). Reading this level from the figure, it is around 860 hPa. Based on the pressure level heights shown on the CD-ROM figure, the actual height of the LCL is roughly

$$969 + 0.4 \times (1992 - 969) \approx 1378\,\text{m}$$

A check of the National Weather Service records for this date and location gives an LCL of 851 hPa or 1456 m, and hence our graphical calculation is reasonably accurate.

If ascent and adiabatic cooling continue, the temperature may drop below 0 °C. This is known as the freezing level, and above this level, ice crystals can start to form in a process known as *glaciation*. In Figure 11.2, the freezing level is at 580 hPa, which corresponds to 4680 m. The process of freezing leads to a release of latent heat, so that continued cooling with ascent will be reduced further. Above the freezing level, even though the cloud is saturated with respect to water, it is *supersaturated* with respect to ice. Deposition of water vapor occurs on the surface of the ice crystals, until humidity is reduced sufficiently that saturation with respect to ice is achieved. This process also releases latent heat.

11.1.1 Stability in the moist atmosphere

Probably the most important driver of uplift for severe weather is moist convection. The stability of a moist atmosphere to a vertical displacement can be understood as follows. Recall that a dry atmosphere is neutrally stable if the potential temperature is uniform with height. Such a dry adiabatic profile corresponds to a particular *lapse rate* (that is, a value of $-dT/dz$) that is usually written as Γ_d and has a constant value of g/c_p, where g is the acceleration due to gravity and c_p is the specific heat of dry air at constant pressure. It is this rate of temperature decrease which is represented by the dry adiabat lines on the skew T–log P diagram in Figure 11.2. If the potential temperature decreases with height, a situation which corresponds to a lapse rate greater than the dry adiabatic lapse rate (that is, the atmosphere cools more quickly than an unsaturated rising air parcel that has a constant potential temperature), the air is unstable to a vertical displacement. An air parcel that is displaced upward any distance vertically will be warmer than the surrounding environmental air, and will experience an upward-directed buoyancy force. Thus, this air parcel will accelerate upward, in the direction of the initial displacement. In such an environment, any vertical disturbance of the atmosphere will tend to grow, with the air accelerating in the direction of the initial displacement.

In a moist atmosphere, stability depends on the vertical profile of the wet-bulb potential temperature (θ_w). Details on the calculation of the wet-bulb potential temperature can be found in Rogers and Yau (1989). A uniform profile of the wet-bulb potential temperature corresponds to the *moist adiabatic lapse rate* Γ_s, in which the temperature decreases less rapidly than the dry case because of the latent heat release from condensation. The moist adiabatic lapse rate is indicated on skew T–log P diagrams by the moist adiabats. If the actual lapse rate in the atmosphere falls between the dry and the moist adiabatic lapse rates, the air will be stable in the dry case, but unstable in the moist case, provided that the vertical displacement is sufficient to saturate the air parcel and reduce its rate of cooling to the moist adiabatic lapse rate. The atmosphere in this case is referred to as *conditionally unstable*, with the condition for instability being whether or not the ascending air becomes saturated. Hence, the presence of moisture in the air adds an additional potential for positive buoyancy, and hence convection, to be generated.

Example Graphically determine the temperature and dew point temperature of an air parcel lifted from the surface to 200 hPa at Fort Worth, Texas at 00 UTC 5 Jun 2005 using the skew T–log P diagram.

In the previous example, we determined the change in air parcel temperature and dew point temperature between the surface and the LCL at 860 hPa. We found the temperature of the rising air parcel by drawing a line starting at the surface temperature that was parallel to the dry adiabats. The dew point temperature of the air parcel was found by drawing a line starting at the surface dew point temperature that was parallel to the mixing ratio lines.

Once the air parcel is lifted above the lifting condensation level the potential temperature and mixing ratio will no longer be conserved. The potential temperature will increase due to latent heat release as water vapor condenses, and the change in air parcel temperature will now be given by the moist adiabats on the skew T–$\log P$ diagram. The mixing ratio will decrease due to the condensation of water vapor, and the dew point temperature will remain equal to the air temperature (that is, the rising air parcel will remain saturated). Thus the dew point temperature of the rising air parcel will also be given by the moist adiabats on the diagram. Graphically this is determined by drawing a line, starting at the LCL, that is parallel to the moist adiabats. This is illustrated on the CD-ROM.

At 200 hPa the temperature and dew point temperature of the air parcel (*not* the environmental air, which is indicated by the profiles) are both equal to $-50\,°C$.

A convenient measure of the energy available to accelerate a parcel vertically is the *convective available potential energy*, or CAPE, which is the parcel buoyancy vertically integrated between the level of free convection (when the potential temperature of the parcel is in excess of the environment) and the level of neutral buoyancy (where this ceases to be the case). An example of this calculation performed graphically for the previous example is shown on the CD-ROM. A useful rule of thumb in evaluating CAPE is that the addition of $1\,g\,kg^{-1}$ of water vapor causes the same increase in CAPE as a warming of $2.5\,°C$. Values of CAPE over $2500\,J\,kg^{-1}$ are very unstable.

CAPE represents the potential for convective motion, but it can only be released if a parcel is initially lifted to the level of free convection. Hence, CAPE may be considered formally to be

$$\mathrm{CAPE} = g \int_{z_{LFC}}^{z_{EL}} \left(\frac{T_{parcel} - T_{env}}{T_{env}} \right) dz \tag{11.7}$$

where z_{EL} is the height of the equilibrium level; that is, the level at which the temperature of a buoyantly rising air parcel is equal to the environmental temperature. In fact, there is usually an energy barrier to be surmounted before this can occur, corresponding to the work needed to raise a parcel to its level of free convection. In fact, large values of this so-called *convective inhibition* often correspond with large values of CAPE, since it is the convective inhibition that allows CAPE above to persist and develop.

Given that CAPE is a measure of the energy available to accelerate a parcel vertically we can estimate the maximum vertical velocity of an air parcel as

$$w = (2 \cdot \mathrm{CAPE})^{0.5} \tag{11.8}$$

This relationship between CAPE and w indicates that for highly unstable environments, with CAPE > 2500 J kg^{-1}, the vertical velocity can reach values in excess of 70 m s^{-1}. In reality, vertical velocities in thunderstorms are often less than this, due to mixing of environmental air with the thunderstorm updraft which reduces the CAPE from the value given by Equation (11.7).

11.2 Air mass thunderstorms

As we have seen, the lifting of air causes adiabatic cooling and can result in the rising air becoming saturated, leading to condensation and latent heat release. If the atmosphere is conditionally unstable, the rising saturated air parcel will be warmer than the environment and the result is a convective cloud referred to as a *cumulus* cloud. As discussed in the previous section, the updraft velocities in convective clouds can be quite large, resulting in the presence of supercooled (that is, below 0 °C) liquid water well above the freezing level. There can be substantial horizontal *entrainment* of environmental air into the cloud, altering the temperature and mixing ratio of the rising air from the expected moist adiabatic values indicated on a skew T–log P diagram. For this cloud to grow into a thunderstorm, continued ascent must occur, allowing the cloud to grow vertically until a highly stable layer (often the tropopause) prevents further lifting and creates an anvil (Figure 11.3). Sometimes updraft velocities are sufficiently vigorous that the cloud tops may overshoot the stable layer. Since the strength of the updraft increases with increasing CAPE, as

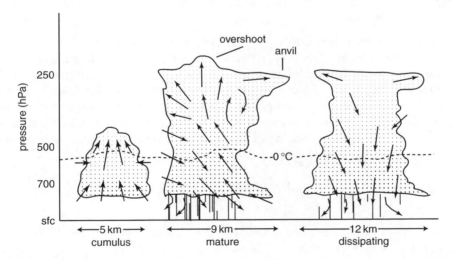

Figure 11.3 Life cycle of a typical air mass thunderstorm cell, showing cumulus, mature, and dissipating stages. Vertical lines between cloud base and the surface (sfc) indicate precipitation (rain or hail). Arrows indicate downdrafts and updrafts within the cloud. The top of the cloud is at the tropopause

shown by Equation (11.8), the presence of an overshooting top is a visual indication of an environment with large CAPE and the potential for severe weather.

At the mature stage of an air mass thunderstorm's life cycle, precipitation will be occurring. A vigorous downdraft is initiated from the downward frictional force exerted on the air by the falling raindrops, and the downdraft region is associated with the heaviest precipitation. At this stage the cloud is called a *cumulonimbus*, or precipitating cumulus, cloud. Severe weather can be associated with these clouds, including rain and strong winds, thunder and lightning, hail and even tornadoes. The maximum updraft velocities in the mature thunderstorm cell are in the center of the cloud. As precipitation continues to develop throughout the cloud and the downdrafts become more extensive, the thunderstorm enters the dissipating stage of its life cycle as the supply of supersaturated air from the lower atmosphere is depleted and the thunderstorm starts to dissipate.

The most common and least severe type of thunderstorm is the air mass thunderstorm. These are single-cell thunderstorms (one cloud with one primary vertical circulation) that develop in maritime tropical air masses, usually in the late afternoons during summer, when surface heating and the consequent convection is at its maximum. Because there is no wind shear that would prevent the destruction of updrafts by precipitation, air mass thunderstorms generally complete their life cycle over the course of an hour.

Example A cumulus cloud has formed in an environment which has a temperature of $-10°C$ and a specific humidity of $0.7\,g\,kg^{-1}$ at 500 hPa. During a process of lateral entrainment, cloud droplets evaporate into the environmental air. What is the lowest temperature to which the entrained air can be cooled by this process, if there are no updrafts or downdrafts?

As unsaturated environmental air is entrained into the cumulus cloud, the water droplets will evaporate. This evaporation will cool the air until saturation is reached, and the temperature attained by this process is known as the *wet-bulb temperature*. In this process, the amount of latent heat required to evaporate the water is provided by the air, and as a result the air temperature will decrease. Energy must be conserved; that is, the same amount of energy used to evaporate the water must be extracted from the air. An equation that will allow us to calculate the wet-bulb temperature can be derived (not shown). This equation is

$$T_w = T - \frac{L}{c_p}\left(\frac{\varepsilon}{p}A\exp\left(\frac{-B}{T_w}\right) - r\right)$$

where T_w is the wet-bulb temperature, T is the environmental air temperature, L is the latent heat of vaporization, c_p is the specific heat of air at constant pressure, $\varepsilon = 0.622$ as defined in Section 11.1, p is the air pressure, r is the mixing ratio, and A ($= 2.53 \times 10^{11}$ Pa) and B ($= 5.42 \times 10^3$ K) are constants used to calculate the saturation vapor pressure of the air. For this example all of the variables in the equation are known, except for T_w. T_w appears on both the left and right hand sides of this equation, so to find T_w we must solve this equation *iteratively*. We start with an assumed value for $T_w = T = 263\,\text{K}$ on the right hand side of the equation, and solve for a new value of T_w:

$$T_w = 263 - \frac{2.5 \times 10^6}{1004}\left(\frac{0.622}{50\,000} \cdot 2.53 \times 10^{11} \cdot \exp\left(\frac{-5.42 \times 10^3}{263}\right) - 0.0007\right)$$

$$= 256\,\text{K}$$

We then use this new value of T_w on the right hand side of the equation to get a further refined estimate of T_w:

$$T_w = 263 - \frac{2.5 \times 10^6}{1004}\left(\frac{0.622}{50\,000} \cdot 2.53 \times 10^{11} \cdot \exp\left(\frac{-5.42 \times 10^3}{256}\right) - 0.0007\right)$$

$$= 260\,\text{K}$$

We repeat (iterate) this process until the values of T_w do not change significantly. For this example, this requires 12 iterations and the final value of T_w is 259 K. This is the temperature to which the entrained air will be cooled and the air will now be saturated.

If this saturated parcel is carried to the 850 hPa level in a downdraft, what will its temperature be?

If the saturated air parcel is forced to sink it will warm adiabatically, and become unsaturated immediately, thus warming at the dry adiabatic lapse rate. The air parcel will conserve potential temperature (Equation (3.9)) and the actual temperature at 850 hPa can be determined from the potential temperature:

$$\theta = T_{500}\left(\frac{p_0}{p_{500}}\right)^{R_d/c_p}$$

$$= 259\left(\frac{1000 \times 10^2}{500 \times 10^2}\right)^{287/1004}$$

$$= 316\,\text{K}$$

$$T_{850} = \theta\left(\frac{p_0}{p_{850}}\right)^{-R_d/c_p}$$

$$= 302\,\text{K}$$

11.3 Multi-cell thunderstorms

In any air mass thunderstorm, the leading edge of the cold air associated with the downdraft is called the *gust front*, which is characterized by a zone of gusty winds

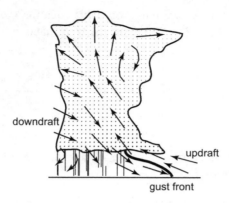

Figure 11.4 A gust front (bold line) between the warm, moist updraft and cool, dry downdraft allows the generation of new cells

and separates the warm moist updraft from the cooler downdraft and associated precipitation (Figure 11.4). Hence, the gust front is able to force the warm, moist air upward, generating a new cell. When a storm continuously regenerates in this manner, a series of single-cell thunderstorms result that are in different stages of development. Such a phenomenon is called a multi-cell thunderstorm. Multi-cell thunderstorms can sometimes be severe.

11.4 Supercell thunderstorms and tornadoes

While air mass thunderstorms are ubiquitous, more interesting weather arises from the dangerous supercell thunderstorm. Nearly all supercell thunderstorms produce severe weather such as large hail, but only around 30% lead to the development of tornadoes. The supercell is a single-cell thunderstorm, with the distinguishing characteristic that the updraft of the storm is rotating, forming a *mesocyclone*. The term *mesoscale* refers to systems that range in size from tens to hundreds of kilometers. Often mesoscale systems are separated into three classes: meso-alpha (200–2000 km); meso-beta (20–200 km), and meso-gamma (2–20 km.) Supercells, however, are not defined by their size, depth, or violence, but rather by this rotating updraft.

Supercell thunderstorms generally form in an environment that has moderate to high values of CAPE (over $1500\,\mathrm{J\,kg^{-1}}$) and large vertical wind shear (more than $20\,\mathrm{m\,s^{-1}}$ in the layer below 6 km). The vertical shear plays a crucial role in preventing the inbuilt 'self-destruct mechanism' of less severe single-cell thunderstorms, in which precipitation destroys the updrafts, by inducing a circulation that separates the zones of precipitation and of uplift (Figure 11.5). Because supercell thunderstorms are coherent dynamical systems, they propagate in a continuous manner. Hence, supercells can persist for up to 12 hours and may travel great distances.

These storms are still not well understood, due to the obvious difficulties in observing them, but some aspects can be described here.

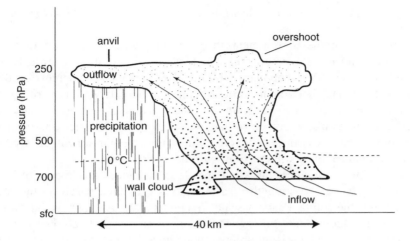

Figure 11.5 Structure of a supercell thunderstorm

A visible manifestation of the mesocyclone is the *wall cloud* (see photograph on the CD-ROM). The development of the mesocyclone arises due to the presence of vertical wind shear, calculated as the vector difference of the wind at two heights. This is because vorticity can be produced wherever there is a gradient in the velocity either normal or parallel to the direction of the existing vorticity vector. When the velocity gradient is normal to the vorticity vector, the lines of vorticity are tilted and new vorticity is generated that is proportional to the steepness of the wind gradient. When the velocity gradient is parallel to the vorticity vector, the lines of vorticity are stretched, generating new vorticity, or compressed, causing a reduction in vorticity.

The vertical wind shear can be enhanced by the presence of a *low-level jet* (a distinct wind maximum in or just above the boundary layer). The shear results in the rotation of air about an axis parallel to the ground. The rotating air is tilted into the vertical by the thunderstorm updraft. The resulting rotation results in lowered pressure in the center of the rotation. This pressure perturbation leads to an even stronger updraft, which in turn enhances rotation through vortex stretching. This process creates a feedback between the strength of the updraft and the cyclonic rotation. The deeper the layer of wind shear, the more efficient this dynamical process. In this context, buoyancy forces are of secondary importance, unlike the weaker air mass thunderstorms. Because of this, supercell thunderstorms can continue after sunset, surviving the loss of surface heating and the resulting buoyancy forces.

Tornado formation is strongly dependent on the dynamical structure of the thunderstorm, with the development of the mesocyclone and associated wall cloud likely to be an essential precursor to tornado formation. The relatively slower (of order $10\,\mathrm{m\,s^{-1}}$) cyclonic circulation of the mesocyclone is generally observed up to an hour before a tornado (spinning with velocities an order of magnitude larger) is evident. It appears that this larger scale circulation leads to the observation that most tornadoes rotate cyclonically, despite the scale argument that tornadoes should have no preferred rotation.

Within this favorable environment, then, tornado formation appears to be related to a smaller, storm-scale process in which warm environmental air rises on the warm side of the outflow boundary while cold air sinks and undercuts on the cold side, which generates horizontal vorticity along the boundary. This vorticity then is tilted and rapidly accelerated vertically into the dynamically generated mesocyclone updraft, resulting in likely tornado formation. The horizontal vorticity associated with this low-level process is in general not evident in the environment. It is generated through the thunderstorm's interaction with the environment.

Because of their violence, the wind speeds in tornadoes are impossible to measure directly. Hence, the strength of a tornado is estimated based on the resulting damage, using the Fujita–Pearson scale (Section 6.2.2). Another useful tool is the weather radar, which cannot observe individual tornadoes, but can discern the associated mesocyclone, which is an order of magnitude larger. Doppler radars can also determine aspects of the wind velocity (see the projects section of the CD-ROM).

11.5 Mesoscale convective systems

In addition to supercell thunderstorms, the atmosphere can generate severe weather in other forms, collectively known as *mesoscale convective systems*. Mesoscale convective systems include everything from individual squall lines to the larger mesoscale convective complexes, and range widely in their degree of organization. There is no strict size definition for these systems, although most often they fall into the meso-beta scale range.

A squall line, or multi-cell line, is a line of thunderstorms which may be simple air mass thunderstorms or supercells that share a common lifting mechanism. Examples of possible lifting mechanisms include a cold front or a *dryline* boundary, which may trigger a rather continuous line of thunderstorms, or a gravity wave, which can generate a more scattered distribution. An example of an active squall line that generated tornadoes in southern Canada and the northern United States along a cold front on 11 July 2005 is shown on the CD-ROM. A dryline is simply the boundary between a moist and a dry air mass across which the temperature does not vary significantly. As with individual thunderstorms, a sufficient environmental CAPE is favorable to squall line development. Severe squall line development is almost always associated with higher values of CAPE (over $2000 \, J \, kg^{-1}$). In addition, a squall line can be self-sustaining in an environment of wind shear, since this allows the individual convective cells to be more severe and longer lived, and perhaps develop into supercells.

Squall lines, like individual thunderstorms, generally undergo a characteristic life cycle. Typically, the squall line develops between 100 and 300 km ahead of and parallel to a cold front, at a location of optimum lift, CAPE, and shear. This location is in the warm, moist sector of the associated mid-latitude cyclone, and the low-level warm advection in this sector contributes strongly to the instability of the environment. Squall line motion is a result of both advection and propagation. Individual thunderstorm cells that make up the squall line tend to move with the low-level wind field, but new cells continue to develop in the direction of the low-level wind shear vector. During the early stages of a severe system, supercells may be present along the entire squall line. However, the interacting circulations of the associated mesocyclones quickly disrupt the line so that only certain locations, such as the ends or at breaks in a squall line, remain favorable for supercell maintenance. In the interior of the squall line, more linear features such as *bow echoes* remain. Whether the narrow band of convective cells develops into a severe system or not, the natural evolution leads to an eventual spreading and weakening of the system.

Bow echoes are, as may be surmised from their name, a type of mesoscale convective system that was identified with the advent of weather radar. Bow echoes are meso-beta scale boomerang-shaped systems of convective cells that occur either as isolated phenomena or as part of a larger system such as a squall line. During the development of a bow echo, cyclonic and anticyclonic vortices develop at either end of the bow. As the cyclonic vortex becomes dominant, the bow evolves into a comma-shaped system. The 'notch' behind the bow signifies the location of the inflow jet, which can be a source of strong, damaging winds. Supercells are sometimes observed in bow echo systems, or may evolve directly into a bow echo as they decay. Severe bow echoes require very high levels of CAPE – observations have suggested that values larger than $2500 \, \text{J kg}^{-1}$ are necessary.

At the largest, meso-alpha, end of the spectrum are *mesoscale convective complexes* (MCCs). Like other deep convective systems, MCCs are usually initiated in regions of high low-level wind shear and CAPE, often in the vicinity of a stationary front. MCCs are defined according to a set of criteria associated with size (greater than $100\,000 \, \text{km}^2$), shape (elliptical rather than linear), and duration (greater than 6 h) of the thunderstorm complex. Typically, MCCs are around $350\,000 \, \text{km}^2$ (as defined by the area of cloud cover) and last for around 11 h, with marine systems tending to be larger and longer lived than terrestrial systems. Peak rainfall tends to occur early in the development of the system. Unlike the sharp weather boundaries associated with squall lines, MCCs are characterized by more persistent disturbances – long, heavy rainfall events leading to flash floods, for example. More than 20% of extreme summer rainfall in the United States has been attributed to MCCs. Similarly, areas of southern China and the Sahel region in Africa rely on deep convective systems such as these for their growing season rainfall.

Review questions

11.1 (a) Calculate the relative humidity with respect to an ice surface for temperatures from 0°C to −40°C, at 5°C intervals, for air that is saturated with respect to a liquid water surface.

(b) At what temperature is the relative humidity with respect to ice greatest?

11.2 Use the surface weather map shown in Figure 1.7 to determine the vapor pressure, saturation vapor pressure, mixing ratio, saturation mixing ratio, and relative humidity at (a) Denver, Colorado, (b) Fort Worth, Texas, and (c) Nashville, Tennessee. The surface pressure is 828.0 hPa at Denver, 984.1 hPa at Fort Worth, and 989.5 hPa at Nashville.

11.3 (a) What is the pressure at the lifting condensation level for an air parcel that has an initial pressure of 1000 hPa, an initial temperature of 20°C, and an initial dew point temperature of 15°C?

(b) Recalculate the lifting condensation level for an initial dew point temperature of 10°C.

(c) Recalculate the lifting condensation level for an initial dew point temperature of 0°C.

(d) Recalculate the lifting condensation level for an initial dew point temperature of −10°C.

11.4 (a) Using the temperature, dew point temperature, and pressure at the surface at Fort Worth, Texas at 00 UTC 5 Jun 2005 from the first example in Section 11.1, calculate the relative humidity using the equations in this chapter.

(b) How does the calculated value of relative humidity compare to that determined based on the values of mixing ratio and saturation mixing ratio estimated from the skew T–log P diagram in the example from Section 11.1?

11.5 Use the storm of 2003 case study data on the CD-ROM from Fort Worth, Texas at 00 UTC 15 Feb 2003 to:

(a) Plot a sounding on a blank skew T–log P diagram using data from the surface and 850, 700, 500, 300, and 250 hPa levels.

(b) Indicate the temperature and dew point temperature of an air parcel lifted from the surface to 250 hPa on the skew T–log P diagram.

(c) Mark the lifting condensation level, level of free convection, and equilibrium level on this sounding.

(d) Shade the portion of the sounding that represents CAPE.

12 Tropical weather

The tropical regions are unique in many ways, and yet the basic dynamics that govern the weather in the tropics is the same as that in the middle latitudes. Nevertheless, phenomena such as tropical cyclones are confined to these low latitudes. Many of the differences observed in tropical dynamics can be attributed to the different scaling that is appropriate there.

12.1 Scales of motion

In the middle latitudes, the strong horizontal temperature gradients and large values of the Coriolis parameter give important guidance as to the appropriate approximations to be made in understanding the dynamics of the weather in these regions. These attributes give rise to the very useful quasi-geostrophic equation, and the conclusion that the available potential energy associated with the strong temperature gradients provides the energy for mid-latitude and polar atmospheric motion. In the tropics, however, both the temperature gradients (Figure 12.1) and the Coriolis parameter

$$f_{45°} = 1.03 \times 10^{-4}$$
$$f_{10°} = 2.53 \times 10^{-5}$$

are small.

Hence, a scale analysis for synoptic scale systems in the tropics may yield rather different results compared to an analysis for the middle latitudes (Section 5.4.2). Consider the basic Navier–Stokes equations with synoptic scales as defined in Table 5.1, so that

$$L \sim 10^6 \, \text{m}$$
$$T \sim \text{several days} \sim 10^5 \, \text{s}$$
$$\Rightarrow U \sim 10 \, \text{m s}^{-1}$$

Applied Atmospheric Dynamics Amanda H. Lynch, John J. Cassano
© 2006 John Wiley & Sons, Ltd

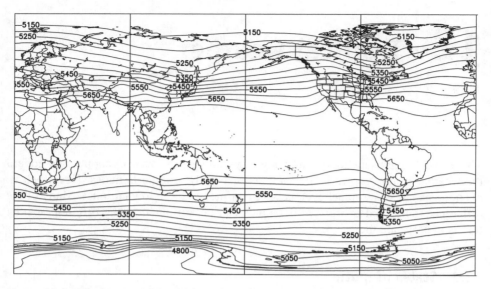

Figure 12.1 The 1000–500 hPa thickness averaged over 1970 to 1999. NCEP Reanalysis data provided by the NOAA-CIRES Climate Diagnostics Center, Boulder, Colorado, from its Web site at http://www.cdc.noaa.gov/

In this case, the Rossby number is still sufficiently small to assume a strong role for the Coriolis force in synoptic scale disturbances:

$$Ro \sim \frac{U}{fL}$$

$$\sim \frac{10}{2.5 \times 10^{-5} \times 10^6}$$

$$Ro \sim 0.4$$

but clearly not as strong as for the mid-latitude case. Assuming that vertical motions are not significantly larger than in the mid-latitude case (this assumption can be justified based on an analysis of the thermodynamic equation), the following scaling can be developed using the dynamic pressure (p_d) form of the equations:

x-eqn	$\frac{\partial u}{\partial t}$	$u\frac{\partial u}{\partial x}$	$v\frac{\partial u}{\partial y}$	$w\frac{\partial u}{\partial z}$	$=-\frac{1}{\rho}\frac{\partial p_d}{\partial x}$	$+v\left(\frac{\partial^2 u}{\partial x^2}+\frac{\partial^2 u}{\partial y^2}\right)$	$+v\frac{\partial^2 u}{\partial z^2}$	$+2\Omega v \sin\phi$	$-2\Omega w \cos\phi$
y-eqn	$\frac{\partial v}{\partial t}$	$u\frac{\partial v}{\partial x}$	$v\frac{\partial v}{\partial y}$	$w\frac{\partial v}{\partial z}$	$=-\frac{1}{\rho}\frac{\partial p_d}{\partial y}$	$+v\left(\frac{\partial^2 v}{\partial x^2}+\frac{\partial^2 v}{\partial y^2}\right)$	$+v\frac{\partial^2 v}{\partial z^2}$	$-2\Omega u \sin\phi$	
scale	$\frac{U^2}{L}$	$\frac{U^2}{L}$		$\frac{UW}{H}$	$\frac{\delta p}{\rho L}$	$\frac{vU}{L^2}$	$\frac{vU}{H^2}$	$f_0 U$	$\approx f_0 W$
magnitude	10^{-4}	10^{-4}		10^{-5}	?	10^{-16}	10^{-12}	2.5×10^{-4}	10^{-6}

The Coriolis term is larger than the time rate of change and advection terms, just as in the middle latitudes, but by a factor of 2, rather than by a factor of 10. In order for

SCALES OF MOTION

these equations to balance, therefore, the pressure gradient term must also be smaller than in the mid-latitude scaling. Hence, the pressure perturbations associated with synoptic scale disturbances must be smaller than for systems of a similar scale in the middle latitudes, and increasingly so as we move toward the equator. There are a range of waves in the tropical atmosphere that conform to this scaling. Here, one important example will be considered.

Example Find the wave equation for a synoptic scale equatorial wave with no meridional velocity perturbation component.

To solve for a wave, the perturbation method is appropriate. For simplicity, we will assume that density is constant. From the scaling developed above

$$\frac{\partial u}{\partial t} + u\frac{\partial u}{\partial x} + v\frac{\partial u}{\partial y} = -\frac{1}{\rho}\frac{\partial p_d}{\partial x} + fv$$

$$\frac{\partial v}{\partial t} + u\frac{\partial v}{\partial x} + v\frac{\partial v}{\partial y} = -\frac{1}{\rho}\frac{\partial p_d}{\partial y} - fu$$

the equations can be linearized assuming a zero basic state ($\bar{u}, \bar{v} = 0$):

$$\frac{\partial u'}{\partial t} = -\frac{1}{\rho}\frac{\partial p'_d}{\partial x} + fv'$$

$$\frac{\partial v'}{\partial t} = -\frac{1}{\rho}\frac{\partial p'_d}{\partial y} - fu'$$

For the equatorial zone, the beta approximation for the Coriolis parameter is simply $f \approx \beta y$, since $f_{0°} = 0$. With no meridional velocity perturbation, the equation set simplifies to

$$\frac{\partial u'}{\partial t} = -\frac{1}{\rho}\frac{\partial p'_d}{\partial x}$$

$$0 = -\frac{1}{\rho}\frac{\partial p'_d}{\partial y} - \beta y u'$$

(12.1)

Hence, this wave will conform to a geostrophic force balance in the meridional direction between the pressure gradient and the zonal velocity perturbation. Assume a solution in the form of zonally propagating waves:

$$u' = \hat{u}(y)\cos(kx - vt)$$

$$p'_d = \hat{p}_d(y)\cos(kx - vt)$$

and substitute into Equation (12.1), resulting in

$$v\hat{u}\sin(kx - vt) = \frac{k}{\rho}\hat{p}_d\sin(kx - vt)$$

$$0 = -\frac{1}{\rho}\frac{\partial \hat{p}_d}{\partial y}\cos(kx - vt) - \beta y \hat{u}\cos(kx - vt)$$

These equations can be combined to eliminate the pressure term and yield a single differential equation for $\hat{u}$:

$$\beta y \hat{u} = -\frac{\nu}{k}\frac{\partial \hat{u}}{\partial y}$$

which can be integrated to give

$$-\frac{\beta k y^2}{2\nu} = \ln \hat{u} + A$$

where A is some constant of integration. If the perturbation zonal velocity at the equator is defined as $\hat{u}_0 \cos(kx - \nu t)$, this constant can be eliminated and the resulting solution is

$$\hat{u} = \hat{u}_0 \exp\left(-\frac{\beta k y^2}{2\nu}\right)$$

$$u' = \hat{u}_0 \exp\left(-\frac{\beta k y^2}{2\nu}\right)\cos(kx - \nu t)$$

Since $c = \nu/k$, this can also be written

$$u' = \hat{u}_0 \exp\left(-\frac{\beta y^2}{2c}\right)\cos(kx - \nu t) \tag{12.2}$$

and the corresponding pressure perturbation is obtained from substituting this solution into either component of Equation (12.1):

$$p'_d = \rho c \hat{u}_0 \exp\left(-\frac{\beta y^2}{2c}\right)\cos(kx - \nu t)$$

In order to yield a solution that decays away from the equator, $c > 0$, and hence these waves are eastward traveling, synoptic, or planetary scale waves trapped near the equator (Figure 12.2). Such waves are known as *Kelvin waves*.

Kelvin waves were first described in 1879 as water waves traveling along a vertical side boundary, by Sir William Thomson, Lord Kelvin (who was introduced in Chapter 1). In their atmospheric, equatorially trapped form, the change of sign of the Coriolis parameter at the equator acts in a manner analogous to a vertical boundary. These waves contribute substantially to the temperature, pressure, and wind variations observed in tropical regions, and are important both for influencing the dehydration of air entering the stratosphere and for the formation of tropical cirrus clouds. Kelvin waves are forced by variations in deep convection.

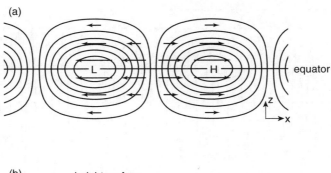
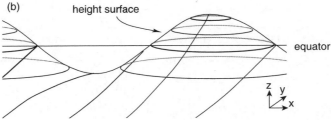

Figure 12.2 Equatorially trapped Kelvin wave shown as (a) a plan view and (b) a three-dimensional view. Vectors in (a) indicate zonal wind associated with the wave. Note that the pressure height surface and zonal wind magnitude vary sinusoidally in x (longitude) and decay exponentially in y (latitude).

12.2 Atmospheric oscillations

Because the atmosphere contains waves of many different wavelengths and frequencies, propagating in many different directions, at all times, it should come as no surprise that these waves sometimes interact to produce oscillating behavior on longer time scales. There are many such *atmospheric oscillations* known in the atmosphere, including the Southern Oscillation (which is associated with El Niño), the Pacific Decadal Oscillation, and the North Atlantic Oscillation. All of these phenomena are characterized by an alternation of two extreme states in a particular region of the globe (characterized by pressure, temperature, and winds), usually with a period on the order of years. For example, a 'warm' El Niño event occurs every 2 to 7 years, and alternates with the 'cold' La Niña state and in between, or 'neutral', states, much as one would expect if taking snapshots of an extremely slow pendulum.

The mechanism driving these oscillations is, in many cases, an open question, but for some an interaction between different waves has been identified as a likely cause. For example, the *quasi-biennial oscillation* or QBO, which has a period of around 26 months, appears to arise from just such a mechanism. This oscillation occurs in the mean zonal winds of the equatorial stratosphere whereby zonally symmetric easterlies and westerlies alternate regularly (Figure 12.3). It is currently thought that the equatorially trapped Kelvin wave provides the westerly momentum and a type of wave known as a combined Rossby–gravity wave provides easterly momentum to

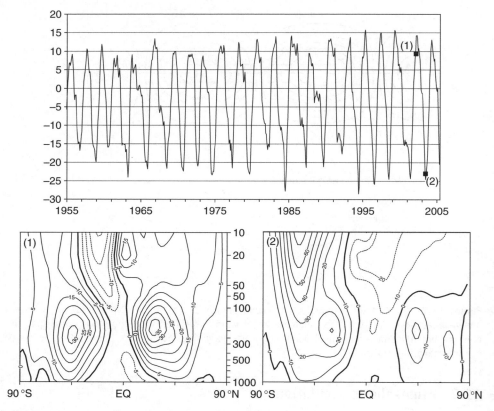

Figure 12.3 The top panel shows the zonal mean zonal wind at the equator at an altitude corresponding to the 30 hPa level, from 1955 to 2005, showing the quasibiennial oscillation between easterlies (negative) and westerlies (positive). An example of the westerly phase is shown in lower panel (1) during March 2002, and of the easterly phase in lower panel (2) during July 2003: these panels show the zonal mean zonal wind by latitude and height. Westerlies are indicated by solid contours and easterlies by dotted contours. Reanalysis data provided by the NOAA-CIRES Climate Diagnostics Center, Boulder, Colorado, from its Web site at http://www.cdc.noaa.gov/

produce the QBO. Recent studies suggest that stratospheric ozone variations are also important to the generation and maintenance of the oscillation.

Although the period of the QBO is nominally defined as a little over 2 years, because it arises from a complex interaction, the period varies substantially over time. The fastest oscillation observed between 1948 and 1993 had a period close to 20 months (1959–1961) and the slowest was 36 months (1984–1987). Each new disturbance appears initially at around 30 km altitude, or around 10 hPa. The disturbance then propagates downward at a rate of around 1 km per month, until it reaches around 23 km (35 hPa). At this point, the westerlies continue downward at the same rate, but the propagation of easterly perturbations is slowed. This asymmetrical descent rate of the easterly and westerly zonal wind perturbations is one of the most notable properties of the QBO.

The QBO is important for many aspects of tropical weather. It is thought to affect rainfall patterns in the Sahel region (the semi-arid transition between the Sahara Desert and the equatorial tropics of Africa). Predictions of both El Niño events and monsoon strength and timing make use of information regarding the phase and strength of the QBO. Finally, the phase of the QBO affects the frequency of tropical cyclones in the Pacific, Atlantic, and Indian Oceans.

12.3 Tropical cyclones

Tropical cyclones are one of the most destructive of meteorological events. They develop over the tropical and subtropical waters of all the world's oceans and are characterized by highly organized convection, a distinct 'eye', and strong cyclonic winds near the surface. In order to be classified as a tropical cyclone, sustained surface winds greater that $33\,\mathrm{m\,s^{-1}}$ or 64 kts must be observed. About 80 tropical cyclones occur each year around the world.

Tropical cyclones go by different names depending on their region of formation. In the north Atlantic Ocean and the eastern side of the Pacific Ocean the term 'hurricane' is used. In the north-west Pacific Ocean the term is 'typhoon', and in other oceanic regions some variant of the more generic 'tropical cyclone' is used.

Tropical cyclones take form in a continuous process, usually spanning several days, from the mesoscale convective complexes (Section 11.5) that develop frequently in the tropical regions. Transition from MCC to tropical cyclone cannot occur unless certain additional preconditions are met in the environment. These requirements are:

1. a conditionally unstable atmosphere (Section 11.1.1);

2. warm ocean temperatures (greater than 26.5 °C);

3. a moist middle troposphere;

4. low vertical wind shear; and

5. at least 500 km from the equator.

The conditionally unstable atmosphere combines with heating from below to allow the development of a robust MCC. Ascent can be initiated from the remains of a higher latitude frontal system, from convergence associated with westward moving disturbances known as African easterly waves, or at the *Intertropical Convergence Zone* (ITCZ) which occurs where the north-easterly and south-easterly trade winds converge. Warm waters at the surface are the ultimate source of energy for the cyclone, and will be described in more detail below. Once convection commences, the moist atmospheric layers in the middle of the troposphere reduce the amount of evaporation within the cloud, and so enhance the rate of precipitation and associated latent heat release. The small wind shear prevents the disruption of the gradually organizing axisymmetric convective rain bands, in contrast to asymmetric mid-latitude systems, which benefit from strong shear.

If the first four preconditions are met, a vigorous convective system will then generate a meso-beta scale vortex, as mid-tropospheric expansion due to heating leads to a lowering of surface pressure and associated surface convergence. This mesoscale circulation, which entails a balance between pressure gradient, friction, and Coriolis force, requires that the development occur some distance from the equator. The thermodynamic disequilibrium between atmosphere and ocean then offsets through evaporation the energy lost to the system due to frictional dissipation. This process is more efficient at higher wind speeds, since strong winds result in very rough ocean conditions with large waves – these evaporate more efficiently than a flat surface or water. Because of this, a feedback can develop, in which escalating winds increase the rate of evaporation, intensifying convection, lowering surface pressure, and further increasing wind speeds. This conceptual model of tropical cyclone intensification is known as WISHE (Wind-Induced Surface Heat Exchange).

When the system reaches wind speeds of between 20 and 34 kts and presents a closed circulation, it is classed as a 'tropical depression'. If winds continue to increase above 34 kts through the feedback process, the system is classed as a 'tropical storm', and then, after 64 kts is reached, a 'tropical cyclone' (Figure 12.4a and b).

Tropical cyclones have an average central pressure of around 960 hPa, much deeper than mid-latitude cyclones. Other differences include the location of strongest winds, which are at the jet stream for mid-latitude cyclones but at the surface for tropical cyclones. Mid-latitude cyclones have cold cores and are asymmetrical in structure, with strong frontal systems. Tropical cyclones have warm cores, no frontal structures, and are relatively axisymmetric. In addition, tropical cyclones are meso-alpha scale in space but of longer duration than mid-latitude mesoscale systems.

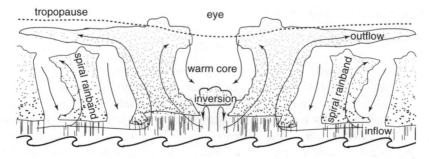

Figure 12.4(a) Cross-section through a mature tropical cyclone, showing the main features and the vertical circulation

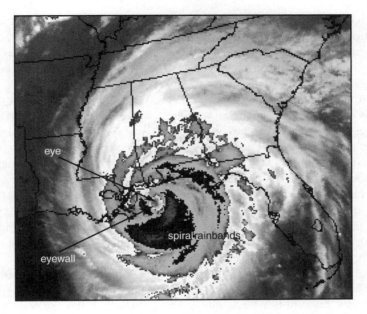

Figure 12.4(b) Key features of a mature topical cyclone, as shown on an enhanced infrared satellite image of Hurricane Ivan, taken at 0315 UTC 16 Sep 2004. Image courtesy of the NOAA National Severe Storms Laboratory, Norman, Oklahoma, USA

Example Use the thermal wind relationship to explain why tropical cyclones are typically characterized by anticyclonic circulation above around 300 hPa.

First, it is useful to write the thermal wind relationship in natural coordinates (see Section 6.4):

$$\frac{\partial V_g}{\partial z} = -\frac{g}{fT_{00}}\frac{\partial T}{\partial n}$$

Since the $\vec{\eta}$ axis is normal to and to the left of the wind, this axis will point to the center of a tropical cyclone in the Northern Hemisphere. Tropical cyclones are warm cored, and hence

$$\frac{\partial T}{\partial n} > 0$$

and the thermal wind relationship requires that

$$\frac{\partial V_g}{\partial z} < 0$$

Hence, the geostrophic wind decreases with height and, if it reverses, will allow an anticyclonic circulation aloft.

A distinctive feature unique to tropical cyclones is the formation of an eye. The eye is an area of relatively light winds (increasing as the eye wall is approached), low pressures, and high temperatures found at the center of most severe tropical cyclones. There is generally little or no precipitation. The eye of a tropical cyclone is usually in the range of 30 to 60 km in diameter.

This warm, clear zone at the center of the cyclone is due to the fact that subsidence within the eye results in adiabatic warming, particularly at middle levels, which also creates an inversion. An inversion profile, as we have seen, is highly stable and hence suppresses convection. The precise mechanisms by which the eye forms are not well understood. One possible constraint that promotes eye formation is the conservation of angular momentum. The winds converging around the center of the cyclone would quickly rise to supersonic speeds to satisfy this requirement, and hence air must ascend before reaching the center. Further, convection in tropical cyclones is organized into long, narrow rain bands that converge toward the center (Figure 12.4), and subsidence on both sides of the bands must contribute to initiating as well as maintaining clearing within the eye.

The eye wall is the area of strong convection, highest surface winds and net uplift. This net upward flow is a result of many updrafts and downdrafts within the individual thunderstorm-like circulations. The formation of the eye wall is related to the convergence of air in a shallow layer above the sea surface. The convergence is due to the disruption of gradient wind balance due to the presence of turbulent stresses in the boundary layer (see Section 6.2.4). The converging air is, of course, very moist and as it rises out of the boundary layer the water vapor condenses to form the eye wall clouds. The air flows outward above the boundary layer and, as a result, clouds in the eye wall tilt outward with height.

Review questions

12.1 Derive Equation (12.1) from the Navier–Stokes equations, listing all assumptions and showing all of the steps in the derivation.

12.2 At 00 UTC 18 Aug (17 Local Standard Time, LST), a Kelvin wave convective envelope at approximately 3°N and 132°W was observed by researchers on a ship. This Kelvin wave had a phase speed of $15\,\mathrm{m\,s^{-1}}$ and a period of 6 days. Plot u' and p'_d for this wave using $\hat{u}_0 = 5\,\mathrm{m\,s^{-1}}$ at $t = 0\,\mathrm{s}$ and $y = 0°$, $1°$, and $5°$ latitude. Describe the phase relationship between u' and p'_d and how these variables vary with latitude.

12.3 Calculate the gradient wind for a landfalling hurricane at Miami, Florida assuming a horizontal pressure gradient of $2\,\mathrm{mb\,km^{-1}}$ at a distance of 50 km from the center of the storm.

12.4 Using the answer from question 12.3 calculate the altitude at which the winds shift from cyclonic to anticyclonic assuming that the eye is 5°C warmer than a location 10 km from the center of the storm.

13 Mountain weather

Many interesting features of mountain weather are related to the generation of internal gravity waves by flow over *orography* (mountains). These waves are commonplace in a stable atmosphere and can be generated by many mechanisms. In this chapter we will derive the governing equations for internal gravity waves, explore how these waves are generated by flow over orography, and how the waves affect the winds, pressure, and temperature to result in some unique aspects of mountain weather. We will conclude the chapter by considering the dynamics associated with severe downslope windstorms that occur in the lee of major mountain barriers around the world.

13.1 Internal gravity waves

13.1.1 Derivation of the internal gravity wave equation

In a stable atmosphere, an air parcel that is displaced vertically will experience a buoyancy force that accelerates the air parcel back toward its original position. This buoyancy force is the restoring force for internal gravity waves, first introduced in Section 8.1. In that example, the dispersion relationship, which relates the frequency of the wave, ν, to the horizontal, k, and vertical, m, wavenumbers, and the Brunt–Väisälä frequency, N, was given as $\nu = Nk/|\vec{k}|$. We will now derive the governing equations for this type of wave motion and show that the dispersion relationship assumed in Section 8.1 is indeed correct.

For our analysis of internal gravity waves we will start with the Navier–Stokes equations (4.19). These waves are very much smaller than synoptic scale systems, and hence we cannot use the synoptic scaling to simplify the equations. However, we can make some simplifications. These include making the Boussinesq approximation (Section 6.3), and ignoring the effects of friction. We will also assume that the flow is two-dimensional in the xz plane, such that we can neglect the meridional component, v, of the wind. Finally we note that the gravity waves are characterized by sufficiently short time and space scales such that the Rossby number is large and the Coriolis force terms can be neglected. With these assumptions the Navier–Stokes equations (4.19) simplify to

$$\frac{\partial u}{\partial t} + u\frac{\partial u}{\partial x} + w\frac{\partial u}{\partial z} + \frac{1}{\rho}\frac{\partial p}{\partial x} = 0 \qquad (13.1\text{a})$$

$$\frac{\partial w}{\partial t} + u\frac{\partial w}{\partial x} + w\frac{\partial w}{\partial z} + \frac{1}{\rho}\frac{\partial p}{\partial z} + g = 0 \qquad (13.1\text{b})$$

$$\frac{\partial u}{\partial x} + \frac{\partial w}{\partial z} = 0 \qquad (13.1\text{c})$$

Finally, we can assume that the flow is adiabatic, which results in an equation for the potential temperature:

$$\frac{\partial \theta}{\partial t} + u\frac{\partial \theta}{\partial x} + w\frac{\partial \theta}{\partial z} = 0 \qquad (13.1\text{d})$$

To derive the governing equations for internal gravity waves we will use the perturbation method introduced in Section 8.2. The variables in Equations (13.1) will be assumed to consist of a basic state and a small departure from that basic state as follows:

$$u = \bar{u} + u'$$
$$w = w'$$
$$p = \bar{p}(z) + p' \qquad (13.2)$$
$$\theta = \bar{\theta}(z) + \theta'$$
$$\rho = \bar{\rho} + \rho'$$

The basic state is constant in both the x and z directions and with time except for $\bar{p}(z)$ and $\bar{\theta}(z)$, which vary in the vertical direction, and we have also assumed that the basic state vertical component of the wind, w, is zero. Further, we will require that the basic state pressure and density satisfy the hydrostatic equation

$$\frac{d\bar{p}}{dz} = -\bar{\rho}g \qquad (13.3)$$

Combining the definition of the basic and perturbation states given by Equations (13.2) and the horizontal momentum equation given by Equation (13.1a) we get

$$\frac{\partial \bar{u}}{\partial t} + \frac{\partial u'}{\partial t} + \bar{u}\frac{\partial \bar{u}}{\partial x} + \bar{u}\frac{\partial u'}{\partial x} + u'\frac{\partial \bar{u}}{\partial x} + u'\frac{\partial u'}{\partial x} + w'\frac{\partial \bar{u}}{\partial z} + w'\frac{\partial u'}{\partial z} + \frac{1}{\bar{\rho}+\rho'}\left(\frac{\partial \bar{p}}{\partial x} + \frac{\partial p'}{\partial x}\right) = 0$$

INTERNAL GRAVITY WAVES

Derivatives of the basic state variables are zero (except for vertical derivatives of $\bar{p}$ and $\bar{\theta}$) and terms that involve products of perturbations are sufficiently small that they can be neglected. This then gives

$$\frac{\partial u'}{\partial t} + \bar{u}\frac{\partial u'}{\partial x} + \frac{1}{\bar{p}+p'}\frac{\partial p'}{\partial x} = 0$$

By noting that

$$\frac{1}{\bar{\rho}+\rho'} = \frac{1}{\bar{\rho}}\left(1+\frac{\rho'}{\bar{\rho}}\right)^{-1} \approx \frac{1}{\bar{\rho}}\left(1-\frac{\rho'}{\bar{\rho}}\right)$$

then for $|\rho'/\bar{\rho}| \ll 1$, this reduces to

$$\frac{\partial u'}{\partial t} + \bar{u}\frac{\partial u'}{\partial x} + \frac{1}{\bar{\rho}}\frac{\partial p'}{\partial x} = 0 \qquad (13.4a)$$

It can be shown that $-\rho'/\bar{\rho} \approx \theta'/\bar{\theta}$. Using this approximation, the other equations (13.1) can be simplified in a similar manner:

$$\frac{\partial w'}{\partial t} + \bar{u}\frac{\partial w'}{\partial t} - \frac{\theta'}{\bar{\theta}}g + \frac{1}{\bar{\rho}}\frac{\partial p'}{\partial z} = 0 \qquad (13.4b)$$

$$\frac{\partial u'}{\partial x} + \frac{\partial w'}{\partial z} = 0 \qquad (13.4c)$$

$$\frac{\partial \theta'}{\partial t} + \bar{u}\frac{\partial \theta'}{\partial x} + w'\frac{\partial \theta'}{\partial z} = 0 \qquad (13.4d)$$

In Equations (13.4) we have four equations and four unknowns (u', w', p', and θ'), if we assume that the basic state is known. To solve this, we can reduce this set to a single equation with a single unknown (the details of this are shown on the CD-ROM) to give

$$\left(\frac{\partial}{\partial t} + \bar{u}\frac{\partial}{\partial x}\right)^2 \left(\frac{\partial^2 w'}{\partial x^2} + \frac{\partial^2 w'}{\partial z^2}\right) + N^2\frac{\partial^2 w'}{\partial x^2} = 0 \qquad (13.5)$$

This is a complex equation, but can be solved rather easily if we assume a form for the solution.

13.1.2 Dispersion relation for internal gravity waves

Like Rossby waves and Kelvin waves, the dispersion relation (Section 8.1) for gravity waves can be derived by assuming a wave-like solution of Equation (13.5):

$$w' = w_0 \cos(kx + mz - \nu t) \qquad (13.6)$$

where w_0 is the amplitude of the wave, k and m are the horizontal and vertical wavenumbers, and ν is the wave frequency. Substitution of this assumed solution into Equation (13.5) results in

$$(\nu - \bar{u}k)^2(k^2 + m^2) - N^2 k^2 = 0 \tag{13.7}$$

which is the dispersion relationship for internal gravity waves. For the case of $\bar{u} = 0$ we find that

$$\nu = \pm \frac{Nk}{\sqrt{k^2 + m^2}}$$

$$= \pm \frac{Nk}{|\vec{k}|} \tag{13.8}$$

which is the equation used in the example in Section 8.1. The dispersion relation tells us that in order for our assumed solution to be valid, this specific relationship between ν, k, m, and N must be satisfied, and arbitrary values of these variables are not allowed. Also, note that, unlike the simplified example in Section 8.1, Equation (13.8) has two roots. The correct root of this equation is selected by noting that frequency is positive (by convention), so for $k > 0$ the positive root is used and for $k < 0$ the negative root is used.

The components of the phase speed, as we saw in Section 8.1, are

$$c_x = \frac{\nu}{k} = \pm \frac{N}{|\vec{k}|}$$

$$c_z = \frac{\nu}{m} = \pm \frac{Nk}{m|\vec{k}|} \tag{13.9}$$

and the components of the group velocity are

$$c_{gx} = \frac{\partial \nu}{\partial k} = \pm \frac{Nm^2}{|\vec{k}|^3}$$

$$c_{gz} = \frac{\partial \nu}{\partial m} = \pm \frac{-Nkm}{|\vec{k}|^3} \tag{13.10}$$

and we choose the appropriate root according to the sign of k, just as for Equation (13.8).

Table 13.1 shows the phase speeds and group velocities resulting from all possible combinations of signs of k and m when $\bar{u} = 0$. The signs of c_x and k are always the same and the signs of c_z and m are always the same, so the wave propagates in the direction of the wavenumber vector, $\vec{k} = (k, m)$, as expected. One interesting

Table 13.1 All possible combinations of phase speed (c_x and c_z) and group velocity (c_{gx} and c_{gz}) are shown here for the case of zero basic wind ($\bar{u}=0$) according to the sign of k (which determines the sign of root chosen) and the sign of m

Sign of k	Sign of m	Root	c_x	c_z	c_{gx}	c_{gz}
+	+	+	+	+	+	−
+	−	+	+	−	+	+
−	+	−	−	+	−	−
−	−	−	−	−	−	+

feature of gravity waves is that c_{gz} is always of opposite sign to c_z; that is, the direction of vertical *energy* propagation in an internal gravity wave is *opposite* the direction of vertical *wave* propagation. Thus, a downward propagating internal gravity wave will direct energy upward and vice versa. This feature will be essential in our understanding of the orographically forced internal gravity waves discussed below.

Example Show that for an internal gravity wave the direction of energy propagation, as given by the group velocity, is perpendicular to the wavenumber vector.

First recall that when the scalar product of two vectors is equal to zero the two vectors are perpendicular (Section 2.2.3). So it is simply necessary to verify that $\vec{k} \bullet \vec{c}_g$ is identically zero:

$$\vec{k} \bullet \vec{c}_g = \left(k\vec{x}+m\vec{z}\right) \bullet \left(c_{gx}\vec{x}+c_{gz}\vec{z}\right)$$

$$= \left(k\vec{x}+m\vec{z}\right) \bullet \left(\frac{Nm^2}{\left|\vec{k}\right|^3}\vec{x} - \frac{Nkm}{\left|\vec{k}\right|^3}\vec{z}\right)$$

$$= \frac{Nkm^2}{\left|\vec{k}\right|^3} - \frac{Nkm^2}{\left|\vec{k}\right|^3}$$

$$= 0$$

And hence the wavenumber and group velocity vectors are perpendicular.

For the case of $\bar{u} \neq 0$, Equation (13.7) gives a more complex relationship:

$$\nu = \pm \frac{Nk}{\left|\vec{k}\right|} + \bar{u}k$$

We can also define an *intrinsic frequency*, which is characteristic of a particular stability profile (or *stratification*) of the atmosphere, as given by N:

$$\hat{\nu} \equiv \nu - \bar{u}k = \pm \frac{Nk}{|\vec{k}|} \tag{13.11}$$

This is the frequency that would be measured by an observer who is moving with the basic wind at a speed of $\bar{u}$. The positive root of this equation applies to cases when $\nu > \bar{u}k$ and the waves are propagating toward the east relative to the mean wind, while the negative root applies to cases when $\nu < \bar{u}k$ and the waves are propagating toward the west relative to the mean wind.

The intrinsic phase speed and group velocity of the wave are identical to Equations (13.9) and (13.10), while the actual phase speed is

$$c_x = \frac{\nu}{k} = \pm \frac{N}{|\vec{k}|} + \bar{u}$$

$$c_z = \frac{\nu}{m} = \pm \frac{Nk}{m|\vec{k}|} \tag{13.12}$$

and the components of the group velocity are

$$c_{gx} = \frac{\partial \nu}{\partial k} = \pm \frac{Nm^2}{|\vec{k}|^3} + \bar{u}$$

$$c_{gz} = \frac{\partial \nu}{\partial m} = \pm \frac{-Nkm}{|\vec{k}|^3} \tag{13.13}$$

As for the dispersion relation, the positive and negative roots of Equations (13.12) and (13.13) apply to the cases of $\nu > \bar{u}k$ and $\nu < \bar{u}k$ respectively. Hence, the only difference between the phase speed and group velocity of an internal gravity wave in the presence of zero or nonzero mean wind is that the mean horizontal wind acts to modify the phase speed and group velocity in the x direction, while leaving the vertical phase speed and group velocity unchanged.

13.1.3 Structure of internal gravity waves

We can now use the equations we have developed to create a diagram of the structure of an internal gravity wave. First, we will use our assumed solution of Equation (13.5) to derive the corresponding solutions for u', p', and θ'.

Example Derive an equation for u' that is not a function of any other perturbation variables.

INTERNAL GRAVITY WAVES

Using the assumed solution for w, Equation (13.6), in Equation (13.4c) gives

$$\frac{\partial u'}{\partial x} + \frac{\partial}{\partial z}[w_0 \cos(kx + mz - \nu t)] = 0$$

$$\frac{\partial u'}{\partial x} = -mw_0 \sin(kx + mz - \nu t)$$

$$u' = -w_0 \frac{m}{k} \cos \phi \qquad (13.14\text{a})$$

So, for m and k of the same sign u' is exactly out of phase with w'; that is, when w' is a maximum u' is a minimum.

Derivation of the equations for p' and θ' are left as review questions. When we assume zero basic flow, the equations result in

$$p' = -\rho w_0 \nu \frac{m}{k^2} \cos \phi \qquad (13.14\text{b})$$

$$\theta' = \frac{w_0}{\nu} \frac{d\bar{\theta}}{dz} \sin \phi \qquad (13.14\text{c})$$

We can now use these equations to take a 'snapshot' of the wind, pressure, and temperature distribution of an internal gravity wave. We will consider the case of $k < 0$ and $m < 0$, an internal gravity wave that is propagating toward the west and downward. The lines of constant phase ϕ at a particular time can be determined by

$$\phi = kx + mz - \nu t|_{\text{constant}} = \text{constant}$$

$$\Rightarrow z = -\frac{k}{m} x + \text{constant} \qquad (13.15)$$

where the constant in Equation (13.15) depends on both the phase and the time. The lines of constant phase for a wave at $t = 0$ s are shown in Figure 13.1. The wavenumber vector, $\vec{k}$, is directed down and toward the left in this figure since $m < 0$ and $k < 0$, and this is the direction of phase propagation. Since the wave is propagating toward the left in this figure the phase decreases toward the right and increases toward the left. Both Figure 13.1 and Table 13.2 indicate the location of maximum, minimum, and zero values of u', w', p', and θ' calculated from Equations (13.6) and (13.14).

Figure 13.1 shows that the air parcel oscillations in the wave are parallel to the wave fronts (lines of constant phase) and perpendicular to the direction of wave

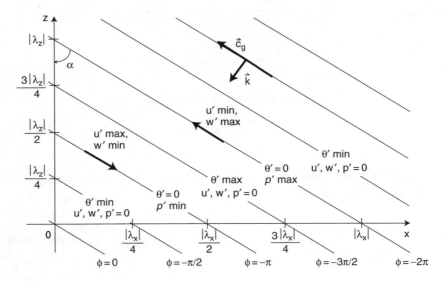

Figure 13.1 Schematic illustration of the phase relationships between u', w', p', and θ' for an internal gravity wave. Locations of maximum, minimum, and zero values of u', w', p', and θ' are labeled. The thin diagonal lines show lines of constant phase (labeled with values of ϕ). The angle between the lines of constant phase and the vertical direction is α (top left corner). Distances along the x and z axes are labeled relative to the horizontal, λ_x, and vertical, λ_z, wavelength of the wave

Table 13.2 Phase location of the maximum, minimum, and zero values of u', w', p', and θ' for an internal gravity wave with $k < 0$, $m < 0$, and $\bar{u} = 0$

ϕ	x	w'	u'	p'	θ'
0	0	Maximum	Minimum	Maximum	Zero
$-\pi/2$	$\|\lambda_x\|/4$	Zero	Zero	Zero	Minimum
$-\pi$	$\|\lambda_x\|/2$	Minimum	Maximum	Minimum	Zero
$-3/2\pi$	$3\|\lambda_x\|/4$	Zero	Zero	Zero	Maximum
-2π	$\|\lambda_x\|$	Maximum	Minimum	Maximum	Zero

propagation. The angle between the air parcel oscillations and the vertical direction, α, is given by

$$\cos \alpha = \frac{\lambda_z}{\left(\lambda_x^2 + \lambda_z^2\right)^{0.5}}$$

$$= \frac{2\pi/m}{\left[(2\pi/k)^2 + (2\pi/m)^2\right]^{0.5}}$$

$$= \frac{2\pi/m}{(2\pi/km)\sqrt{k^2 + m^2}}$$

$$= \frac{k}{\sqrt{k^2+m^2}}$$

$$\cos\alpha = \frac{k}{|\vec{k}|}$$

Combining this result with the dispersion relationship, Equation (13.8), gives

$$\nu = N\cos\alpha \tag{13.16}$$

So, the frequency of the wave depends only upon the Brunt–Väisälä frequency and the angle of the air parcel oscillations in the wave. For air parcel oscillations that are vertical, $\alpha=0$, the wave propagates in the horizontal direction, and the frequency of the wave is equal to the Brunt–Väisälä frequency ($\nu=N$). As the angle of air parcel oscillation becomes increasingly horizontal, the frequency of the wave decreases, eventually to a value of 0 for air parcel oscillations that are purely horizontal. More rapid air parcel oscillations, with a frequency greater than N, are not possible in internal gravity waves.

13.2 Flow over mountains

13.2.1 Idealized case: an infinite series of ridges

Armed with the knowledge of internal gravity wave dynamics we now turn our attention to the generation of such waves in response to flow over mountains. We will first consider the simplest, most idealized case: a constant mean wind flowing over an infinite series of sinusoidal ridges. This orography can be written

$$h(x) = \bar{h} + h'$$
$$= h'$$
$$= h_0\cos(kx)$$

where we have assumed that the orography consists of a mean and perturbation part and that the mean orography ($\bar{h}$) has zero elevation. The perturbation orography consists of an infinite series of ridges with a horizontal wavenumber of k and a height of h_0.

Since the wind cannot flow into or out of the ground, the flow at the surface must be parallel to the orography. Hence, the perturbation vertical velocity at the ground will be given by

$$w'(x,0) = \frac{dh}{dt}$$
$$= \bar{u}\frac{\partial h'}{\partial x}$$
$$w'(x,0) = -\bar{u}kh_0\sin(kx) \tag{13.17}$$

Consider the case of a stationary wave: $\partial/\partial t = 0$ and $\nu = 0$. The governing equation for internal gravity waves (Equation (13.5)) with these assumptions simplifies to

$$\bar{u}^2 \frac{\partial^2}{\partial x^2}\left(\frac{\partial^2 w'}{\partial x^2} + \frac{\partial^2 w'}{\partial z^2}\right) + N^2 \frac{\partial^2 w'}{\partial x^2} = 0$$

Integrating this equation twice with respect to x gives

$$\frac{\partial^2 w'}{\partial x^2} + \frac{\partial^2 w'}{\partial z^2} + \frac{N^2}{\bar{u}^2} w' = 0 \tag{13.18}$$

Substituting into Equation (13.18) an assumed wave-like solution of the form $w' = w_0 \sin(kx + mz)$ (where we have used $\sin$ to be congruent with the boundary condition in Equation (13.17)) gives the dispersion relationship

$$m^2 = \frac{N^2}{\bar{u}^2} - k^2$$

$$m = \pm\sqrt{\frac{N^2}{\bar{u}^2} - k^2} \tag{13.19}$$

Applying the lower boundary condition to the assumed solution gives

$$w'(x, z) = -\bar{u}kh_0 \sin(kx + mz) \tag{13.20a}$$

Since the stationary wave assumption requires that $\nu = 0$, Equation (13.11) gives $\hat{\nu} = -\bar{u}k$. This gives an intrinsic phase speed of the wave of

$$\hat{c}_x = \frac{\hat{\nu}}{k} = \frac{-\bar{u}k}{k} = -\bar{u}$$

That is, the wave has a phase speed exactly opposite to the mean wind. Because the wave propagation and the advection by the mean wind exactly cancel, the wave remains stationary relative to the ground, and hence this is consistent with our assumption.

We can determine the sign of m in Equation (13.19) by noting that the vertical group velocity (Equation (13.13)), and thus vertical energy propagation, is in the opposite direction to the vertical wave propagation (Equation (13.12)). For the case of flow over orography, the source of wave energy is at the lower boundary, and so energy must be propagating upward, and therefore $m < 0$.

The structure of the wave is shown in Figure 13.2(a) for the case of $\bar{u} > 0$. The lines of constant phase tilt westward with height. Relative to the mean wind, the wave propagates down and to the west and energy propagates up and to the west as shown by the group velocity vectors on this figure.

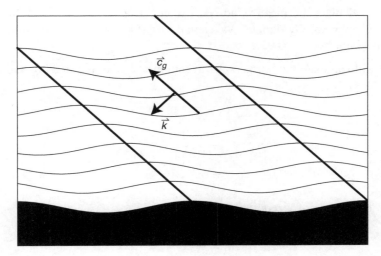

Figure 13.2(a) Vertical cross-section through a vertically propagating internal gravity wave generated by flow over an infinite series of ridges. Streamlines are shown as thin lines, and the mean flow is from left to right in the figure. Streamlines are plotted at intervals of an eighth of a vertical wavelength, and the top streamline is plotted at a height of one vertical wavelength. The heavy black lines mark lines of constant phase. The group velocity vector ($\vec{c}_g$) and the wavenumber vector ($\vec{k}$) are also shown

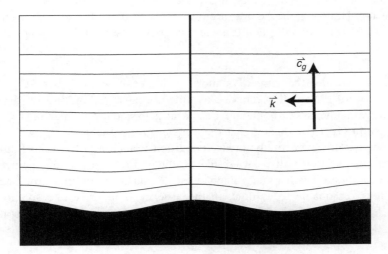

Figure 13.2(b) Vertical cross-section through an evanescent internal gravity wave generated by flow over an infinite series of ridges. Streamlines are shown as thin lines, and the mean flow is from left to right in the figure. Streamlines are plotted at intervals of an eighth of a vertical wavelength, and the top streamline is plotted at a height of one vertical wavelength. The heavy black line indicates a line of constant phase. The group velocity vector ($\vec{c}_g$) and the wavenumber vector ($\vec{k}$) are also shown

Example Calculate the mean wind speed that causes $N^2/\bar{u}^2 = k^2$ for flow over a series of ridges with a horizontal wavelength of 10 km when the atmospheric static stability is given by $N = 0.02\,\text{s}^{-1}$.

First, we will calculate the horizontal wavenumber, k, that is forced by these ridges:

$$k = \frac{2\pi}{\lambda_x}$$
$$= \frac{2\pi}{10\,000}$$
$$= 6.3 \times 10^{-4}$$

Then, the magnitude of $\bar{u}$ required for $N^2/\bar{u}^2 = k^2$ is

$$|\bar{u}| = \frac{N}{|k|}$$
$$= \frac{0.02}{6.3 \times 10^{-4}}$$
$$= 32\,\text{m s}^{-1}$$

This wind speed is often referred to as the *critical wind speed*. When the magnitude of the mean flow is less than this critical wind speed, Equation (13.19) gives a real result. However, if the mean flow is faster than this critical wind speed, Equation (13.19) is not valid and it is clear we have assumed the wrong form of the solution.

Let us now consider the case of a more rapid mean wind. When $|\bar{u}| > N/|k|$ a different form of the solution must be assumed because the forcing for the internal gravity wave at the lower boundary occurs at a frequency greater than N, and as such Equation (13.18) indicates that the atmosphere is unable to support a wave at this frequency. However, the boundary condition Equation (13.17) must still apply, and hence the solution will be wave-like in the x direction and this component of the assumed solution will remain unchanged. Because the forcing originates at the ground, it is reasonable to assume, in the absence of a wave-like solution in the vertical, a bounded solution in the z direction. Hence, we assume a solution of the form

$$w' = w_0 e^{-mz} \sin(kx) \tag{13.20b}$$

which satisfies Equation (13.17) if $w_0 = -\bar{u}kh_0$ as before. A bounded solution requires that $m > 0$. Verification that this solution also satisfies the governing Equation (13.18) is left as a review question. The structure of the solution for this case is shown in Figure 13.2(b).

FLOW OVER MOUNTAINS

The case of the vertically propagating internal gravity wave (Figure 13.2a) is known as the wide ridge case since widely spaced ridges will have a small wavenumber k, and hence are likely to satisfy the condition $|\bar{u}k| < N$. This case is also favored for situations with weak winds (small $\bar{u}$) or strong static stability (large N). Conversely, the case of the vertically decaying internal gravity wave, also known as an *evanescent wave*, is referred to as the narrow ridge case, where the ridges are closely spaced and k is large. This case is also favored when the mean wind speed is large or the static stability is weak.

13.2.2 Flow over an isolated ridge

Flow over an infinite series of sinusoidal ridges, while illustrative of the processes acting to generate topographic internal gravity waves, is not encountered in the real world. Instead, the atmospheric flow is more likely to encounter isolated topographic obstacles. Fortunately, the methods used above can also be used to determine the structure and characteristics of internal gravity waves generated by flow over an isolated ridge.

The topography of an isolated ridge, $h'(x)$, can be represented as a sum of sines and cosines, with varying amplitudes, h_n, and wavenumbers, k_n, of the same form as that used to represent an infinite series of ridges. Summing a large number of sines and cosines over a range of appropriately chosen wavelengths can create a single curve that approximates the shape of the isolated ridge.

The atmospheric response to flow over each individual sine or cosine term is identical to that discussed in the previous section (Equation (13.20)). For $|\bar{u}k_n| < N$ Equation 13.20(a) applies and the wave is vertically propagating, while for $|\bar{u}k_n| > N$ Equation 13.20(b) applies and the wave is evanescent. The perturbation method yields a linear equation, and hence we can sum the atmospheric response for the flow over each individual sine or cosine term to reconstruct the resulting wave field. For wide ridges the sum is dominated by sines and cosines with small k_n and the resulting wave is vertically propagating as illustrated in Figure 13.3(a). For narrow ridges the sum is dominated by sines and cosines with large k_n and the wave that is generated by the flow over this topography is evanescent. An example of this type of situation is shown in Figure 13.3(b).

Example Minna Bluff is a ridge 5 km wide and 900 m tall that extends onto the Ross Ice Shelf of Antarctica. This ridge is oriented east/west while the prevailing flow in this region is from the south. Observations during a recent winter found a wind of 20 m s^{-1} from the south and $N = 0.02$ s^{-1}. Determine if the flow over Minna Bluff for these conditions will result in a vertically propagating or evanescent internal gravity wave.

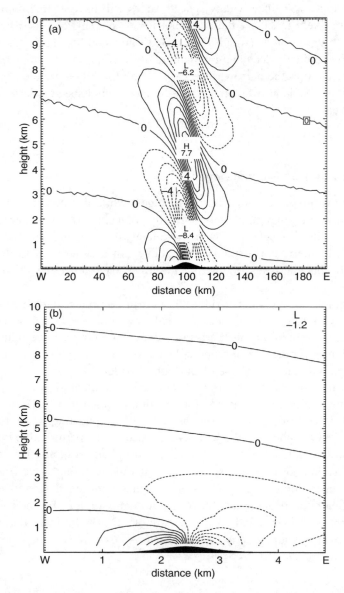

Figure 13.3 Vertical cross-section through (a) a vertically propagating internal gravity wave and (b) an evenescent internal gravity wave generated by flow over an isolated ridge. Contours of vertical velocity are plotted, with upward vertical velocity shown by solid contour lines and downward vertical velocity shown by dashed lines. The ridge is shown at the bottom of the figure by the solid shading, and is shown with a vertical exaggeration of two times

We calculate the horizontal wavenumber of this ridge as

$$k = \frac{2\pi}{\lambda_x}$$
$$= \frac{2\pi}{5000}$$
$$= 1.3 \times 10^{-3}\,\mathrm{m}^{-1}$$

and note that the representation of this ridge as a sum of sines and cosines will be dominated by sines and cosines with a wavenumber greater than $1.3 \times 10^{-3}\,\mathrm{m}^{-1}$ (that is, wavelengths less than 5 km).

We use this value of k to evaluate $|\bar{u}k|$ as

$$|\bar{u}k| = 20 \times 1.3 \times 10^{-3}$$
$$= 0.026\,\mathrm{s}^{-1}$$

and note that this value of $|\bar{u}k|$ is greater than N. Therefore this flow will result in an evanescent internal gravity wave of the type illustrated in Figure 13.3(b).

13.2.3 Flow with vertical variations in wind speed and stability

While Section 13.2.2 addressed the more realistic case of internal gravity waves generated by flow over an isolated ridge, the solution for this type of flow is still limited by the assumption of constant background wind speed and stability. In the real atmosphere these conditions are rarely met, and if they are present are likely to represent a rather uninteresting example. More typical is the case where the background wind, $\bar{u}$, and the static stability, N, vary in the vertical direction.

For this more general case the governing equation for the internal gravity waves, Equation (13.18), contains an additional term proportional to the curvature of the wind profile:

$$\frac{\partial^2 w'}{\partial x^2} + \frac{\partial^2 w'}{\partial z^2} + \frac{N^2}{\bar{u}^2}w' - \bar{u}^{-1}\frac{d^2\bar{u}}{dz^2}w' = 0 \qquad (13.21)$$

where the coefficients of w' are grouped as a single term known as the *Scorer parameter*:

$$l^2 = \frac{N^2}{\bar{u}^2} - \bar{u}^{-1}\frac{d^2\bar{u}}{dz^2} \qquad (13.22)$$

In fact, often the second term is small and can be neglected.

As in Section 13.2.1 we will assume a wave-like solution of the form $w' = w_0 \sin(kx + mz)$. Substituting this assumed solution into Equation (13.21) gives the dispersion relationship

$$m^2 = l^2 - k^2$$

$$m = \pm\sqrt{l^2 - k^2}$$

Our assumed solution is valid for $l^2 > k^2$ and results in a vertically propagating wave. If $l^2 < k^2$ our assumed solution is not valid, and instead we must assume a solution of the form $w' = w_0 e^{-mz} \sin(kx)$, which results in an evanescent wave. We can define a critical horizontal wavenumber as $k_c = l$, where internal gravity waves will be vertically propagating for $k < k_c$ and evanescent for $k > k_c$.

Let us now consider an atmosphere with two layers, each with constant l, such that in the lower layer $l_L > k_c$ and in the upper layer $l_U < k_c$. This suggests an atmosphere which is highly stable with comparatively weak winds in the lower layer and lower stability, perhaps combined with strong winds, in the upper layer.

Flow over an isolated ridge in this situation will generate a vertically propagating internal gravity wave in the lower layer. This vertically propagating wave will be reflected at the boundary between the two layers, since vertical wave propagation is

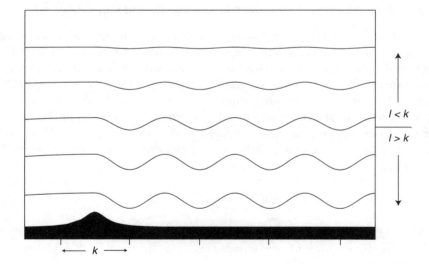

Figure 13.4 Vertical cross-section through a trapped lee wave generated by flow over an isolated ridge. Streamlines are shown as thin lines, and the mean flow is from left to right in the figure. Tick marks along the horizontal axis indicate the horizontal wavelength of the trapped lee wave and of the topography. The tick mark along the right side of the figure indicates the boundary between layers of the atmosphere with differing Scorer parameter (l). The relationship between l and k in each layer is also indicated

not possible in the upper layer. The wave will then be repeatedly reflected between the interface between the two atmospheric layers and the ground surface, resulting in a *trapped lee wave* downstream of the isolated ridge. This situation is illustrated schematically in Figure 13.4. If sufficient moisture is present in the atmosphere clouds will form at the crest of each wave shown in Figure 13.4. Clouds that form in this way are known as *lee wave clouds* and are a common sight downwind of mountainous regions, particularly during the winter months.

Example Lee wave clouds were observed over Boulder, Colorado on 20 December 2004. Use the rawinsonde observations listed in Table 13.3 from Grand Junction, Colorado (upstream of the Rocky Mountains) at 12 UTC 20 Dec 2004 to determine the horizontal scale of topography required to generate trapped lee waves.

Trapped lee waves require that $l_L > k_c$ and $l_U < k_c$. We will calculate l for the lower layer (surface to 700 mb) and upper layer (700 mb to 300 mb) using Equation (13.22) and neglecting the curvature term. We first calculate $\bar{u}, \bar{\theta}$, and N for each layer as

$$\bar{u}_L = \frac{15.9 + (-1.5)}{2} \qquad \bar{u}_U = \frac{24.1 + 15.9}{2}$$
$$= 7.2 \text{ m s}^{-1} \qquad = 20.0 \text{ m s}^{-1}$$

$$\bar{\theta}_L = \frac{300.8 + 281.7}{2} \qquad \bar{\theta}_U = \frac{320.1 + 300.8}{2}$$
$$= 291.3 \text{ K} \qquad = 310.5 \text{ K}$$

$$N^2 = \frac{g}{\bar{\theta}} \frac{d\theta}{dz}$$

$$N_L^2 = \frac{9.81}{291.3} \times \frac{(300.8 - 281.7)}{(3053 - 1453)} \qquad N_U^2 = \frac{9.81}{310.5} \times \frac{(320.1 - 300.8)}{(9270 - 3053)}$$
$$= 4.02 \times 10^{-4} \text{ s}^{-2} \qquad = 9.81 \times 10^{-5} \text{ s}^{-2}$$

Table 13.3 Rawinsonde observations from Grand Junction, Colorado at 12 UTC 20 Dec 2004

Pressure (mb)	Elevation (m)	Potential temperature (K)	Zonal component of the wind ($m s^{-1}$)
850	1453	281.7	−1.5
700	3053	300.8	15.9
300	9270	320.1	24.1

Using these values we calculate the Scorer parameter:

$$l^2 = \frac{N^2}{\bar{u}^2}$$

$$l_L^2 = \frac{4.02 \times 10^{-4}}{7.2^2}$$
$$= 7.7 \times 10^{-6} \text{ m}^{-2}$$
$$\Rightarrow l_L = 2.8 \times 10^{-3} \text{ m}^{-1}$$

$$l_U^2 = \frac{9.81 \times 10^{-5}}{20^2}$$
$$= 2.5 \times 10^{-7} \text{ m}^{-2}$$
$$\Rightarrow l_U = 4.9 \times 10^{-4} \text{ m}^{-1}$$

The wavenumber of the orography that is forcing the wave must be less than l_L, but greater than l_U. The horizontal wavelength of the topography is given by

$$\lambda_x = \frac{2\pi}{k}$$

and hence the bounds of orographic wavelength are derived from the critical wavelength in each layer so that

$$\lambda_{cL} = \frac{2\pi}{2.8 \times 10^{-3}}$$
$$= 2243 \text{ m}$$
$$= 2.2 \text{ km}$$

$$\lambda_{cU} = \frac{2\pi}{4.9 \times 10^{-4}}$$
$$= 12\,823 \text{ m}$$
$$= 12.8 \text{ km}$$

Hence, for a trapped lee wave, the orography must have a wavelength longer than 2.2 km and less than 12.8 km.

The width of the Colorado Rocky Mountains is approximately 200 km, too large to satisfy the conditions identified above for trapped lee waves. Within the Colorado Rocky Mountains are numerous smaller mountain ranges, with the Front Range being located just to the west of Boulder, Colorado. The horizontal scale of the high peaks in the Front Range is roughly 6 km, and could force a trapped lee wave for the atmospheric conditions observed on 20 December 2004.

As has been shown in this section, internal gravity waves generated by flow over topography can take many forms depending on the scale of the topographic forcing (k) and the state of the atmosphere ($\bar{u}$, N, l). We now turn our attention to the case of strong downslope windstorms which can result in property damage and loss of life downwind of major mountain ranges in extreme situations.

13.3 Downslope windstorms

The lee slopes of mountain ranges around the world occasionally experience strong, damaging downslope winds. These winds are often a defining characteristic of the local climate and are given unique regional names. In the United States strong downslope winds in the lee of the Rocky Mountains are known as *chinooks* and can have speeds in excess of 60 m s^{-1}. In the Alps the same type of downslope windstorm is known as a *foehn*, while downwind of the Southern Alps of New Zealand these windstorms are referred to as *norwesters*, for the north-west wind direction common to these events in this part of the world.

To understand the dynamics of downslope windstorms, we will consider the flow of a constant density fluid over a mountain ridge, similar to the flow of water over a rock in a stream. To simplify the analysis, we will try to define a height of the fluid surface, given by $h(x)$. If the height of the barrier is given by $h_m(x)$, then the depth of the fluid layer is simply $h - h_m$.

Starting with the x component of the Navier–Stokes equations (4.19), simplifying assumptions can be made. We will consider flow in the xz plane such that we can neglect the meridional component, v, of the wind and spatial derivatives in the y direction ($\partial/\partial y = 0$). We will assume that the zonal component, u, of the flow does not vary in the vertical direction ($\partial u/\partial z = 0$). We will ignore the effects of friction and will only consider flows with sufficiently short time and space scales such that the Coriolis force terms can be neglected. Finally, we will only consider steady state flows such that $\partial/\partial t = 0$. With these assumptions the x component of the Navier–Stokes equations (4.19) simplifies to

$$u\frac{\partial u}{\partial x} = -\frac{1}{\rho}\frac{\partial p}{\partial x}$$

where the density is constant but u and p may vary. Note that this equation, unlike the equations considered in previous sections of this chapter, is a nonlinear equation (see Section 5.4.2), due to the advection term on the left hand side.

We will also require that the flow satisfies the hydrostatic equation

$$\frac{\partial p}{\partial z} = -\rho g$$

We can then use this to rewrite the pressure gradient force term, by assuming that the pressure at the top of the fluid layer is equal to zero. In this case, the pressure at any depth z in the fluid is given by $p(z) = \rho g(h-z)$, and the pressure gradient force term can be written as

$$-\frac{1}{\rho}\frac{\partial p}{\partial x}\bigg|_z = -g\frac{\partial h}{\partial x}$$

This equation states that the horizontal pressure gradient depends only upon the slope of the fluid surface, and further that the pressure gradient will not vary in the vertical direction in the fluid. Using this result the u momentum equation can be written

$$u\frac{\partial u}{\partial x} = -g\frac{\partial h}{\partial x} \tag{13.23}$$

If the height of the fluid surface slopes downward in the positive x direction the zonal component of the flow must increase in this direction. If the fluid surface slopes upward in the positive x direction the zonal component of the flow must decrease in this direction.

Equation (13.23) still contains two unknown variables, u and h. To create a closed set of equations, the principle of continuity can be invoked. For a fluid of constant density,

$$\frac{\partial u}{\partial x} + \frac{\partial w}{\partial z} = 0$$

Integrating this equation over the depth of the fluid, from h_m to h, yields

$$\int_{z=h_m}^{h} \frac{\partial w}{\partial z} \partial z = -\int_{z=h_m}^{h} \frac{\partial u}{\partial x} \partial z$$

$$w(h) - w(h_m) = -(h - h_m)\frac{\partial u}{\partial x}$$

but the vertical velocities at the boundaries of the fluid layer are simply

$$w(h) = u\frac{\partial h}{\partial x} \qquad w(h_m) = u\frac{\partial h_m}{\partial x}$$

and the continuity equation then reduces to

$$u\frac{\partial h}{\partial x} - u\frac{\partial h_m}{\partial x} = -(h - h_m)\frac{\partial u}{\partial x}$$

$$\frac{\partial [u(h - h_m)]}{\partial x} = 0 \qquad (13.24)$$

This equation requires that changes in the zonal wind speed must be compensated for by changes in the depth of the fluid, $h - h_m$, such that as the wind speed increases the total depth of the fluid must decrease and vice versa.

Equations (13.23) and (13.24) now make a closed set of equations in the two unknown variables u and h. These equations can be combined by multiplying Equation (13.23) by u and using Equation (13.24) to replace $\partial h/\partial x$ to give

$$[u^2 - g(h - h_m)]\frac{\partial u}{\partial x} = -ug\frac{\partial h_m}{\partial x}$$

We simplify this equation by defining the Froude number, Fr, as

$$Fr = \frac{u^2}{g(h - h_m)}$$

and hence

$$(1 - Fr^2) \frac{\partial u}{\partial x} = \frac{u}{h - h_m} \frac{\partial h_m}{\partial x} \qquad (13.25)$$

The changes in the zonal flow in response to the topography depend on the value of the Froude number. For $Fr < 1$ the flow is termed *subcritical* and $\partial u/\partial x \propto \partial h_m/\partial x$. When $Fr > 1$, $\partial u/\partial x \propto -\partial h_m/\partial x$ and the flow is termed *supercritical*. The Froude number can be understood as the ratio of the magnitude of nonlinear advection to the magnitude of the pressure gradient. Thus, in supercritical flow, the nonlinear advection term dominates and balance is satisfied only when the flow is accelerated. The flow for both of these cases is illustrated in Figure 13.5.

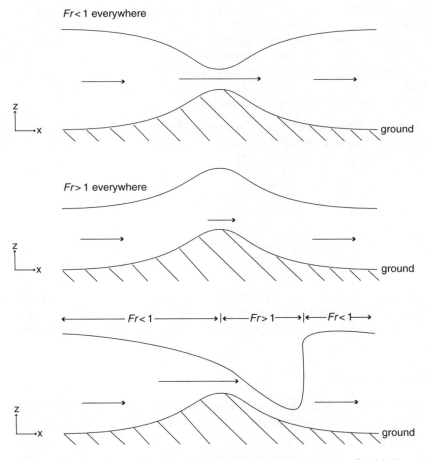

Figure 13.5 Vertical cross-sections of a two-layer fluid flowing over a ridge for (a) $Fr < 1$ everywhere (subcritical flow), (b) $Fr > 1$ everywhere (supercritical flow), and (c) $Fr < 1$ transitioning to $Fr > 1$ over the ridge (hydraulic jump)

Consider a case where a subcritical flow approaches the crest of a mountain. The wind speed increases and the total depth of the fluid layer decreases, both of which cause Fr to increase. If the value of Fr upstream of the barrier is sufficiently large that Fr increases to a value greater than 1, the flow will transition from subcritical to supercritical. As the supercritical flow descends the lee of the barrier the wind speed will increase further and the depth of the fluid will continue to decrease. This flow eventually becomes unstable and a turbulent *hydraulic jump* occurs. The jump allows the flow to adjust back to subcritical conditions: the wind speed decreases and the depth of the fluid increases. This situation can result in very large wind speeds in the lee of the barrier, the *downslope windstorm*. Conditions in the hydraulic jump are extremely turbulent and pose a serious hazard to aircraft.

Example Measurements were taken on 24 January 1992 at experimental towers near the town of Bettles, located in the lee of the Brooks Range in Alaska, near the top of the Brooks Range, and on the northern, windward side. During the early hours of the morning, a rapid and turbulent acceleration of the wind from relatively calm to 90 kts was observed at the Bettles site. Around the same time, the prevailing northerlies were impinging on the Brooks Range at 10 kts. Soundings indicated a very shallow stable layer extending to around 1700 m was topped by a strong inversion. Does the data indicate the presence of a hydraulic jump?

The elevation of the Brooks Range is around 1500 m. Assuming that the altitude of the inversion does not change substantially as the air flows over the Brooks Range, the Froude number can be calculated:

$$Fr_{upstream} = \frac{u^2_{upstream}}{g(h-h_m)} \qquad Fr_{crest} = \frac{u^2_{upstream}}{g(h-h_m)}$$

$$= \frac{5.14^2}{9.81 \times 1700} \qquad = \frac{46.3^2}{9.81 \times 200}$$

$$= 1.6 \times 10^{-3} \qquad = 1.1$$

This calculation indicates that the formation of the strong inversion created a shallow stable layer, which enabled the development of a hydraulic jump as the air flowed over the Brooks Range. This was experienced as a turbulent, strong, but short-lived windstorm in Bettles.

Review questions

13.1 Determine the wind speed required for our assumption of negligible Coriolis terms in the Navier–Stokes equations for flow over the following mountain ranges:

(a) Rocky Mountains, Colorado, width = 200 km

(b) Southern Alps, New Zealand, width = 100 km

(c) Snowy Range, Wyoming, width = 20 km.

13.2 Starting with the Navier–Stokes equations, as simplified by the assumptions at the start of this chapter (Equations (13.1)), use the definition of the basic state and perturbation variables in Equations (13.2) to derive the perturbation governing equations (13.4). (Hint: use the fact that the basic state is assumed to satisfy the hydrostatic assumption and that $-\rho'/\bar{\rho} \approx \theta'/\bar{\theta}$.)

13.3 Verify that the dispersion relationship for internal gravity waves given in Section 13.1.2 is correct.

13.4 Derive equations for p' and θ' for an internal gravity wave that are not functions of any other perturbation variables, such that the phase relationship between u', w', p', and θ' can be determined.

13.5 Sketch a figure similar to Figure 13.1 for $k > 0$ and $m > 0$, showing lines of constant phase, direction of phase propagation ($\vec{k}$), direction of energy propagation ($\vec{c}_g$), and locations of maximum, minimum, and zero u', w', p', and θ', and wind vectors.

13.6 Verify that the dispersion relationship given by Equation (13.19), for internal gravity waves generated by flow over an infinite series of ridges, is correct.

13.7 Verify that the assumed solution given by Equation (13.20b) for an evanescent internal gravity wave is a solution of Equation (13.18).

13.8 Consider the case of a stably stratified flow ($N = 0.02$ s^{-1}) over an infinite series of ridges with $h_0 = 50$ m and a horizontal wavelength of 1 km.

(a) Calculate $u_{critical}$ for this situation.

(b) Plot h as a function of x for $-\lambda_x \leq x \leq \lambda_x$.

(c) Plot w' as a function of x for $-\lambda_x \leq x \leq \lambda_x$ at $z = 0$ and a height equal to one vertical wavelength for $\bar{u} = 1$ and 9 m s^{-1}.

13.9 It is winter in Wyoming, and the flow approaching the Snowy Range ($\lambda_x = 20$ km) has the following properties:

P (mb)	Z (m)	θ (K)
850	1500	270
700	3000	287
500	5500	289

The mean wind speed in the layer between 850 and 700 mb is 5 m s^{-1} and the mean wind speed in the layer between 700 mb and 500 mb is 30 m s^{-1}.

(a) Calculate the Brunt–Väisälä frequency for each layer.

(b) Calculate the Scorer parameter for each layer, neglecting the curvature term.

(c) What types of internal gravity waves would be expected in each layer for this flow over the Snowy Range?

(d) Sketch the flow over and downwind of the Snowy Range for this case.

14 Polar weather

While the dynamics that govern the weather of the polar regions are the same as those that govern the weather of the middle latitudes and tropics, there are some unique aspects of polar weather. One such feature is the juxtaposition of strong low-level static stability with rugged orography. The large ice sheets of Greenland and Antarctica reach elevations in excess of 3000 m and are marked by steep ice slopes along their margins. The weather over these ice sheets is strongly constrained by this confluence of orography and stability, resulting in the characteristic *katabatic winds*. Over the polar oceanic regions intense low-pressure systems, smaller than typical mid-latitude cyclones, occur. These *polar lows* can be as intense as tropical cyclones, and in fact share some common characteristics with their tropical relatives.

14.1 Katabatic winds

A defining characteristic of the atmospheric circulation over the Antarctic and Greenland ice sheets is the persistent low-level wind. This *katabatic wind* is defined as a wind that flows downslope under the influence of gravity. The flow is negatively buoyant; that is, the air in the katabatic flow is colder and thus more dense than the ambient air outside of the flow. Katabatic winds can occur over any sloped surface when the air in contact with that surface is cooled (due to turbulent heat transfer to the underlying surface and/or by direct radiative cooling). Short-lived katabatic winds can occur in mountainous terrain anywhere on Earth on a nightly basis as the air near the surface cools after sunset. Over the large polar ice sheets, persistent inversion conditions (temperature increasing with altitude away from the ice sheet surface, see Section 4.3.4) result in katabatic winds that can last for days, weeks, and even months at a time.

Early explorers of the Antarctic continent were the first to report the existence of strong katabatic winds along the edge of the ice sheet. An expedition led by Sir Douglas Mawson from 1911 to 1914 reported wind speeds at Cape Denison (67.1°S, 142.7°E) of 25 to 35 m s^{-1} (49 to 68 kts) for months at a time, with gusts in excess of 45 m s^{-1}. More recent meteorological observations at Cape Denison, made by automatic weather stations, have reported an annual mean wind speed of 20.9 m s^{-1}

Applied Atmospheric Dynamics   Amanda H. Lynch, John J. Cassano
© 2006 John Wiley & Sons, Ltd

(41 kts) for the year 1995. Annual mean wind speeds from other years are less reliable as the anemometer at this site is often destroyed by the strong katabatic flow long before the harsh Antarctic winter ends and the weather station can be repaired.

Another unique characteristic of the katabatic wind regime is the nearly constant wind direction. The directional constancy (*DC*), defined as the ratio of the vector mean wind speed to the scalar mean wind speed, was an amazing 0.96 at Cape Denison during 1995. A directional constancy of 1.0 would indicate that the wind was always from the same direction. The Cape Denison resultant wind direction (that is, the vector-averaged wind direction), measured counterclockwise from due north, was 162° (SSE). This wind direction is oriented approximately 20° to the left of the local downslope direction of the ice topography, often referred to as the *fall line*. We will see later that over the Antarctic ice sheet we should expect katabatic winds to be directed to the left of the ice sheet fall line, while over the Greenland ice sheet we would expect the katabatic winds to be directed to the right of the ice sheet fall line.

Example Calculate the scalar mean wind speed, resultant wind speed and direction, and the directional constancy for March 1997 at Terra Nova Bay, Antarctica (74.95°S, 163.69°E) using the automatic weather station observations provided on the CD-ROM.

The scalar wind speed is simply the average of the magnitude of the wind speed observations for the month, but the resultant wind speed takes into account both positive and negative values of the wind components u and v. Consider the simple case of two wind observations: one is a wind of $10\,\mathrm{m\,s^{-1}}$ from the south and the other is a wind of $10\,\mathrm{m\,s^{-1}}$ from the north. The scalar average wind speed would simply be $10\,\mathrm{m\,s^{-1}}$. For the resultant wind speed we would average the u and v components of the wind speed and then calculate the total wind speed from the averages of the two components. For the south wind we have $u = 0\,\mathrm{m\,s^{-1}}$ and $v = 10\,\mathrm{m\,s^{-1}}$. For the north wind we have $u = 0\,\mathrm{m\,s^{-1}}$ and $v = -10\,\mathrm{m\,s^{-1}}$. The average u component for these two wind observations would be $0\,\mathrm{m\,s^{-1}}$ and the average v component would also be $0\,\mathrm{m\,s^{-1}}$. The resultant wind speed in this case would be $0\,\mathrm{m\,s^{-1}}$. The scalar wind speed will always be greater than or equal to the resultant wind speed.

The scalar mean wind speed can be obtained from the individual wind speed observations as follows:

$$|\vec{V}| = \frac{1}{N}\sum_{i=1}^{N} WS_i$$

$$= 15.2\ \mathrm{m\,s^{-1}}$$

where $N = 248$ is the number of three-hourly weather observations for the month of March 1997 at this location.

Each of the individual wind speed and direction observations can be converted into zonal and meridional components:

$$u = -WS \times \sin(WD)$$
$$v = -WS \times \cos(WD)$$

where *WS* is the observed wind speed and *WD* is the observed wind direction. Then, the mean zonal and meridional components of the wind are

$$\bar{u} = \frac{1}{N}\sum_{i=1}^{N} u_i$$
$$= 14.4 \text{ m s}^{-1}$$

$$\bar{v} = \frac{1}{N}\sum_{i=1}^{N} v_i$$
$$= -3.3 \text{ m s}^{-1}$$

Using these results we can calculate the resultant wind speed as

$$\text{resultant } WS = \sqrt{\bar{u}^2 + \bar{v}^2}$$
$$= 14.8 \text{ m s}^{-1}$$

The resultant wind direction can be calculated from the horizontal wind components using the formula

$$WD = 90 - \frac{180}{\pi}\tan^{-1}\left(\frac{\bar{v}}{\bar{u}}\right) + WD_0$$

where $WD_0 = 180°$ for $\bar{u} > 0$ and $WD_0 = 0°$ for $\bar{u} < 0$. This equation is not valid for $\bar{u} = 0$: in this case $WD = 180°$ when $\bar{v} > 0$, $WD = 360°$ when $\bar{v} < 0$, and $WD = 0°$ when $\bar{v} = 0$. The resultant wind direction is then

$$\text{resultant } WD = 90 - \frac{180}{\pi}\tan^{-1}\left(\frac{\bar{v}}{\bar{u}}\right) + WD_0$$
$$= 90 - \frac{180}{\pi}\tan^{-1}\left(\frac{-3.3}{14.4}\right) + 180$$
$$= 283$$

The fall line at Terra Nova Bay is directed toward the east, and so the resultant wind direction of 283° is oriented 13° to the left of the fall line.

The directional constancy at Terra Nova Bay is

$$DC = \frac{\sqrt{\bar{u}^2 + \bar{v}^2}}{\frac{1}{N}\sum_{i=1}^{N}|\vec{V}_i|}$$
$$= \frac{\sqrt{14.4^2 + (-3.3)^2}}{15.2}$$
$$= 0.97$$

Strong winds, large directional constancy, and a wind direction that is 10° to 50° to the left of the local fall line are all characteristics of Antarctic katabatic winds. We will now explore the dynamics responsible for driving this persistent wind regime.

Consider the near surface atmospheric state over the Antarctic ice sheet during the long polar night. At a location in the interior of the Antarctic ice sheet, at 75°S, 135°E for example, the ice surface slopes gently to the north. The Sun has set in early May and will not rise again until August. Without sunlight the surface cools by emitting long-wave radiation and the air immediately in contact with the surface cools due to the turbulent transfer of heat to the ice surface. This results in the development of a strong temperature inversion in the lowest 100 to 1000 m of the atmosphere. We will approximate this situation by an atmosphere that consists of two layers, each of constant density, with the lower layer having a greater density than the upper layer. The lower layer is of constant depth, h, and the interface between the lower and upper layers is parallel to the underlying ice topography, as illustrated in Figure 14.1.

For an analysis of katabatic winds the horizontal wind components of the Navier–Stokes equations (4.19) can be approximated as follows. Observations of katabatic winds in the Antarctic indicate that the winds are nearly constant in time, so we will assume that the flow is steady state ($\partial/\partial t = 0$). We will also neglect nonlinear advection terms in the equations of motion, an assumption which is evaluated in review question 14.4 at the end of this chapter. We will orient our coordinate system such that the positive x direction points in the downslope direction and the positive y direction points to the left of the downslope direction. With these

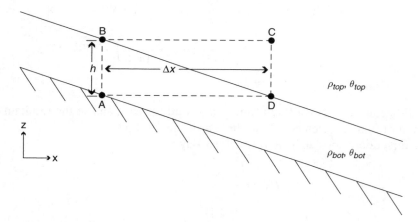

Figure 14.1 Vertical cross-section through the two-layer atmosphere used to represent katabatic winds. h is the depth of the inversion layer, $h/\Delta x$ is the slope of the ice sheet surface, ρ is the density, and θ is the potential temperature. The subscripts bot and top refer to the bottom and top layers respectively. The points labeled A, B, C, and D are discussed in the text

assumptions and using Equations (10.4) to simplify the friction term, Equations (4.19) simplify to

$$0 = -\frac{1}{\rho}\frac{\partial p}{\partial x} + fv - C_D V u$$

$$0 = -\frac{1}{\rho}\frac{\partial p}{\partial y} - fu - C_D V v \qquad (14.1)$$

where V is the scalar wind speed. Using the hydrostatic equation, the pressure at point A (p_A) in Figure 14.1 is given by

$$\frac{\partial p}{\partial z} = -\rho g$$

$$\Rightarrow \int_{p_A}^{p_B} \partial p = \int_{z_A}^{z_B} -\rho_{bot} g \partial z$$

$$p_B - p_A = -\rho_{bot} g (z_B - z_A)$$

$$p_A = p_B + \rho_{bot} g (z_B - z_A)$$

Similarly, the pressure at point D (p_D) in Figure 14.1 is given by

$$p_D = p_C + \rho_{top} g (z_C - z_D)$$

The downslope pressure gradient force acting between points A and D can then be estimated as

$$-\frac{1}{\rho}\frac{\partial p}{\partial x} \approx -\frac{1}{\bar{\rho}}\frac{p_D - p_A}{\Delta x}$$

$$\approx -\frac{1}{\bar{\rho}}\frac{(p_C + \rho_{top} g h) - (p_B + \rho_{bot} g h)}{\Delta x}$$

$$\approx -\frac{1}{\bar{\rho}}\left(\frac{p_C - p_B}{\Delta x} + \frac{(\rho_{top} - \rho_{bot}) g h}{\Delta x}\right)$$

where $\bar{\rho}$ is the mean density in the two layers and Δx is the horizontal distance between points A and D. The first term on the right hand side of this equation, the pressure gradient force in the upper layer, can be written as

$$-\frac{1}{\bar{\rho}}\frac{p_C - p_B}{\Delta x} \approx -\frac{1}{\rho}\frac{\partial p}{\partial x}\bigg|_{top}$$

This term is referred to as the ambient or background pressure gradient force. Because $h/\Delta x$ is simply the terrain slope $-\partial z/\partial x$, the equation for the downslope pressure gradient force is

$$-\frac{1}{\rho}\frac{\partial p}{\partial x} = -\frac{1}{\rho}\frac{\partial p}{\partial x}\bigg|_{top} + g\frac{(\rho_{top} - \rho_{bot})}{\bar{\rho}}\frac{\partial z}{\partial x}$$

Using the fact that

$$\frac{p_{top} - p_{bot}}{\bar{\rho}} \approx -\frac{\theta_{top} - \theta_{bot}}{\bar{\theta}} = -\frac{\Delta\theta}{\bar{\theta}}$$

where $\bar{\theta}$ is the mean potential temperature in the two layers, this equation can be rewritten in terms of the inversion strength $\Delta\theta$

$$-\frac{1}{\rho}\frac{\partial p}{\partial x} = -\frac{1}{\rho}\frac{\partial p}{\partial x}\bigg|_{top} - g\frac{\Delta\theta}{\bar{\theta}}\frac{\partial z}{\partial x}$$

The second term on the right hand side of this equation is referred to as the *katabatic force*, which arises solely due to the presence of a potential temperature inversion over sloping terrain. If the terrain slope is zero or the potential temperature profile is uniform, this term is zero. For a stably stratified atmosphere, where $\theta_{top} > \theta_{bot}$, and for terrain sloping down in the positive x direction, this term is positive and air will be accelerated in the downslope direction.

Since our coordinate system is oriented with the positive x direction aligned with the ice fall line there is no terrain slope in the y direction and the pressure gradient force in the y direction in the lower layer is due only to the ambient pressure gradient force so that

$$-\frac{1}{\rho}\frac{\partial p}{\partial y} = -\frac{1}{\rho}\frac{\partial p}{\partial y}\bigg|_{top}$$

Now, Equations (14.1) can be rewritten as

$$0 = -\frac{1}{\rho}\frac{\partial p}{\partial x}\bigg|_{top} - g\frac{\Delta\theta}{\bar{\theta}}\frac{\partial z}{\partial x} + fv - C_D V u \quad (14.2a)$$

$$0 = -\frac{1}{\rho}\frac{\partial p}{\partial y}\bigg|_{top} - fu - C_D V v \quad (14.2b)$$

If the ambient pressure gradient is neglected, the downslope wind is given by a balance between the katabatic force, the Coriolis force, and the frictional force. We will consider a situation where the katabatic force term is positive: an inversion over terrain that slopes down in the positive x direction. For the simple case of no frictional force, $C_D = 0$, the Coriolis force and the katabatic force create a geostrophic balance. In the Southern Hemisphere this results in a wind that is perpendicular to the fall line of the ice sheet and directed such that higher elevations of the ice sheet are to the left of the wind vector (Figure 14.2a). For the case of no Coriolis force, an unrealistic situation for high latitudes but one which may pertain on a tropical glacier such as the Quelccaya Ice Cap in the Andes Mountains of South America, the downslope momentum equation would be given by a balance between the katabatic force and the frictional force. In this case the wind would be directed down the fall line with

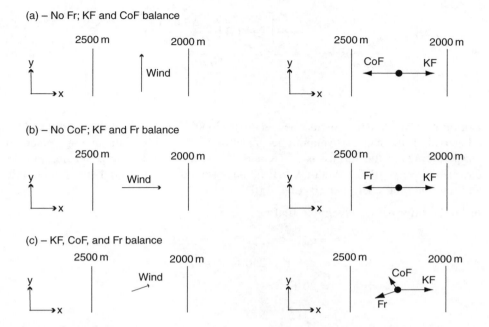

Figure 14.2 Katabatic wind vector (left column) and force vectors for flows when the katabatic and Coriolis forces balance (a), the katabatic and friction forces balance (b), and the katabatic, Coriolis, and friction forces balance (c). Force vectors are shown with filled arrowheads and KF is the katabatic force, CoF is the Coriolis force, and Fr is the frictional force. Wind vectors are shown with small arrowheads. Ice sheet elevation contours are labeled as 2000 m and 2500 m, and in all panels the higher elevation is to the left

no cross-slope component (Figure 14.2b). The more realistic case of all three forces acting results in a wind that is directed downslope and to the left of the fall line in the Southern Hemisphere (Figure 14.2c).

We will now define ψ as the deviation angle of the wind from the terrain fall line (and the x direction), with positive values taken for winds that are directed to the left of the fall line. Using this definition we can write

$$u = V \cos \psi$$
$$v = V \sin \psi$$

Neglecting the ambient pressure gradient force terms in Equations (14.2), V and ψ are given by

$$V = \frac{g}{f} \frac{\Delta \theta}{\overline{\theta}} \frac{\partial z}{\partial x} \sin \psi$$

$$= \left(-\frac{g}{C_D} \frac{\Delta \theta}{\overline{\theta}} \frac{\partial z}{\partial x} \cos \psi \right)^{0.5} \quad (14.3a)$$

$$\cos\psi = \frac{\dfrac{f}{C_D} + \left[\dfrac{f^2}{C_D^2} + 4\left(\dfrac{g}{f}\dfrac{\Delta\theta}{\overline{\theta}}\dfrac{\partial z}{\partial x}\right)^2\right]^{0.5}}{2\left(\dfrac{g}{f}\dfrac{\Delta\theta}{\overline{\theta}}\dfrac{\partial z}{\partial x}\right)} \tag{14.3b}$$

Example Calculate the downslope and cross-slope components of the wind speed for a katabatic flow over the Antarctic ice sheet at 75°S, 135°E if the average potential temperature near the surface is 230 K and 250 K in the middle troposphere, and the terrain slope is −0.002. Assume that the background pressure gradient is negligible and that the drag coefficient is 1.67×10^{-5} m^{-1}.

From the information given, we find

$$\Delta\theta = \theta_{top} - \theta_{bot}$$
$$= 250\text{ K} - 230\text{ K}$$
$$= 20\text{ K}$$

$$\overline{\theta} = \frac{\theta_{top} + \theta_{bot}}{2}$$
$$= \frac{250\text{ K} + 230\text{ K}}{2}$$
$$= 240\text{ K}$$

$$f = 2\Omega\sin\phi$$
$$= 2 \times 7.29 \times 10^{-5}\text{ s}^{-1} \times \sin(-75°)$$
$$= -1.4 \times 10^{-4}\text{ s}^{-1}$$

Using $\partial z/\partial x = -0.002$ and $g = 9.81$ m s^{-1} in Equation (14.3b) gives

$$\cos\psi = \frac{\dfrac{-1.4 \times 10^{-4}}{1.67 \times 10^{-5}} + \left[\left(\dfrac{-1.4 \times 10^{-4}}{1.67 \times 10^{-5}}\right)^2 + 4\left(\dfrac{9.81}{1.4 \times 10^{-4}}\dfrac{20}{240}(-0.002)\right)^2\right]^{0.5}}{2\left(\dfrac{9.81}{-1.4 \times 10^{-4}}\dfrac{20}{240}(-0.002)\right)}$$

$$= 0.70$$

and hence $\psi = 45°$, so the wind is directed at an angle of 45° to the left of the ice slope fall line. We can now use this value in Equation (14.3a) to find the scalar wind speed:

$$V = \left(-\frac{9.81}{1.67 \times 10^{-5}} \times \frac{20}{240} \times (-0.002) \times 0.70\right)^{0.5}$$
$$= 8.3\text{ m s}^{-1}$$

The downslope component of the wind is

$$u = 8.3 \times \sin(45°)$$
$$= 5.9 \text{ m s}^{-1}$$

and the cross-slope component of the wind is

$$v = 8.3 \times \cos(45°)$$
$$= 5.9 \text{ m s}^{-1}$$

Thomas Parish and David Bromwich used this approach to estimate the average katabatic wind flow over the whole of the Antarctic continent (Figure 14.3). We see that the katabatic winds diverge from the high interior of the continent and flow downslope towards the coast as expected.

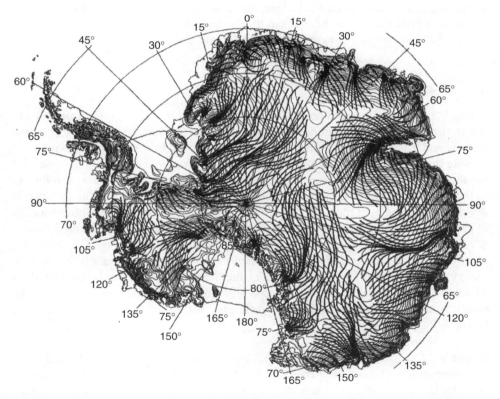

Figure 14.3 Mean wintertime streamlines over the Antarctic continent calculated using the two-layer model of katabatic flow. Ice sheet elevation contours are shown as thin gray lines. (From Parish and Bromwich, 1987. Reprinted with permission of *Nature* Publishing Group)

The persistent katabatic winds over the Antarctic continent, with flow directed downslope away from the high elevations of the ice sheet, are an important component of the meridional circulation of high southern latitudes. This divergence of katabatic winds from the high interior of the Antarctic continent acts to remove mass from over the continent, and results in sinking motion over the continent. As a result of the sinking motion cloud and precipitation formation is inhibited and Antarctica receives very little precipitation. Offshore, the katabatic winds converge with the winds of the middle latitudes, resulting in rising motion, and extensive cloudiness over the Southern Ocean. The final branch of this circulation is a poleward-directed flow in the middle and upper troposphere of the high southern latitudes, forming the polar cell discussed in Chapter 15.

14.2 Barrier winds

The combination of strong static stability and steep topography can result in other types of low-level flows. For example, when a low-level wind in a stable atmosphere is directed toward a topographic barrier, the flow may either pass over the barrier, as discussed in Chapter 13, or be blocked by the barrier. Whether the flow passes over the barrier or is blocked is determined by a Froude number given by

$$Fr = \frac{V}{(gH\Delta\theta/\theta)^{0.5}} \tag{14.4}$$

where V is the wind speed directed toward the barrier, H is the height of the barrier, $\Delta\theta$ is the potential temperature difference between the surface and the top of the barrier, and θ is the potential temperature of the flow approaching the barrier.

Note that the Froude number defined here differs from that defined in Section 13.3 for the discussion of downslope windstorms. This unfortunate situation of referring to these two different quantities both as the Froude number has a long history that is discussed in detail by Baines (1995). For the case of barrier winds the Froude number is proportional to the ratio of the kinetic energy of the flow to the potential energy required for the flow to pass over the barrier. When $Fr > 1$ the flow has sufficient kinetic energy to pass over the barrier. When $Fr < 1$ the flow lacks sufficient kinetic energy to allow the airstream to pass over the barrier and the flow is blocked. This situation is favored when the wind speed is weak or when the atmosphere is very stable, so that the potential temperature difference between the top and bottom of the barrier is large.

Example The Ross Ice Shelf is a large, flat, floating ice mass centered roughly on the International Date Line at the edge of the Antarctic continent. This ice shelf is surrounded by elevated terrain to the east, south, and west, with the steepest terrain being the Transantarctic Mountains that form the western boundary of the ice shelf. Calculate the Froude number for an observed easterly wind of $10\,\mathrm{m\,s^{-1}}$ flowing across the Ross Ice Shelf. The Transantarctic Mountains have an elevation around 2000 m and climatological conditions on the ice shelf during the austral winter give a near surface potential temperature of 260 K and a potential temperature difference between the ice shelf and the top of the Transantarctic Mountains of 10 K.

$$Fr = \frac{10}{(9.81 \times 2000 \times 10/260)^{0.5}} = 0.36$$

Since $Fr < 1$, we would expect this easterly wind across the Ross Ice Shelf to be blocked by the Transantarctic Mountains. What is the atmospheric response to a blocked flow?

Consider a two layer model of the atmosphere as shown in Figure 14.4. The top panel of this figure illustrates the initial conditions, with a stably stratified atmosphere ($\rho_{bot} > \rho_{top}$ and $\theta_{bot} < \theta_{top}$) and a lower layer of undisturbed depth h which is experiencing an easterly flow. As this stable air impinges on the barrier it is unable to flow over the barrier (since $Fr < 1$) and accumulates at the base of the barrier. In response to this situation the lower layer bulges upward adjacent to the barrier, resulting in a modification of the surface pressure at point A in the bottom panel of Figure 14.4. Using the hydrostatic equation, the pressure at point A in the figure is given by

$$p_A = p_B + \rho_{bot}(h + \Delta h)g$$

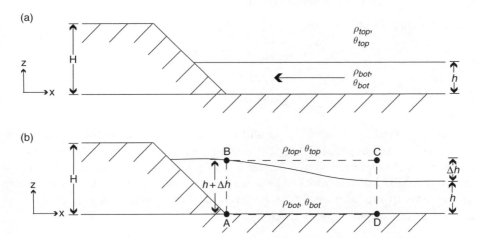

Figure 14.4 Vertical cross-sections of a two-layer atmosphere with the flow in the lower layer impinging upon a barrier (a) and the response of the two-layer fluid to this flow (b). h is the undisturbed depth of the lower layer, Δh is the increased depth of the lower layer adjacent to the barrier, H is the height of the barrier, ρ is the density, and θ is the potential temperature. The subscripts bot and top refer to the bottom and top layers respectively. The points labeled A, B, C, and D are discussed in the text

and the pressure at point D in the figure is given by

$$p_D = p_C + \rho_{bot} hg + \rho_{top} \Delta hg$$

The difference in pressure between points A and D is thus

$$p_A - p_D = p_B + \rho_{bot}(h + \Delta h)g - p_C - \rho_{bot} hg - \rho_{top} \Delta hg$$
$$= p_B - p_C + \rho_{bot} hg - \rho_{bot} hg + \rho_{bot} \Delta hg - \rho_{top} \Delta hg$$
$$= p_B - p_C + (\rho_{bot} - \rho_{top}) \Delta hg$$

If there is no ambient horizontal pressure gradient (that is, $p_B = p_C$), then the pressure difference between points A and D reduces to

$$p_A - p_D = (\rho_{bot} - \rho_{top}) \Delta hg \qquad (14.5)$$

Since $\rho_{bot} > \rho_{top}$ and $\Delta h > 0$, the pressure at point A is greater than the pressure at point D.

Example Calculate the pressure difference between a point adjacent to the Transantarctic Mountains and a point located near the center of the Ross Ice Shelf for the previous example. Assume that the undisturbed depth of the lower layer flow is 500 m and the depth of this layer adjacent to the Transantarctic Mountains is 1000 m. Consistent with the stable stratification, we can take the density of the bottom layer to be $1.3\,kg\,m^{-3}$ and of the upper layer to be $1.2\,kg\,m^{-3}$.

Using Equation (14.5),

$$p_A - p_D = (1.3 - 1.2) \times 500 \times 9.81$$
$$= 491\,\text{Pa}$$
$$= 4.9\,\text{mb}$$

The easterly wind in the lower layer will respond to this new pressure distribution, and come into geostrophic balance. The resulting geostrophic wind will be parallel to the barrier and is referred to as a barrier wind. In order to calculate the geostrophic wind that results from the blocked flow we need to estimate the distance over which the pressure difference occurs. The Rossby radius of deformation (r_R) provides an estimate of this distance and is given by

$$r_R = \frac{1}{|f|} \left(gh \frac{\Delta \theta}{\theta_{bot}} \right)^{0.5} \qquad (14.6)$$

Physically, the Rossby radius of deformation is the horizontal distance over which the atmosphere will adjust to geostrophic balance for a given disturbance, and depends on the intensity of the disturbance and its time scale, here $1/|f|$.

Using the pressure difference given by Equation (14.5) and the horizontal scale of the disturbance suggested by Equation (14.6) we can estimate the horizontal pressure gradient:

$$\frac{\partial p}{\partial x} \approx \frac{\Delta p}{r_R}$$

for flow against a north–south-oriented barrier. The geostrophic wind is then given by

$$v_g = \frac{1}{\rho f} \frac{\partial p}{\partial x}$$

$$v_g \approx \frac{1}{\rho f} \frac{\Delta p}{r_R} \tag{14.7}$$

Example Calculate the geostrophic barrier wind that results from the blocked flow in the previous example. The latitude at the center of the Ross Ice Shelf is 80°S.

At 80°S f is

$$f = 2 \times 7.29 \times 10^{-5} \cdot \sin(-80°)$$
$$= -1.4 \times 10^{-4} \text{ s}^{-1}$$

The Rossby radius of deformation can be calculated using Equation (14.6) using $\Delta\theta = 10\,K$, $\theta_{bot} = 260\,K$, and $h = 500\,$m (the undisturbed depth of the lower layer) from the previous example, and $f = 1.4 \times 10^{-4} \text{s}^{-1}$:

$$r_R = \frac{1}{|-1.4 \times 10^{-4}|} \left(9.81 \times 500 \times \frac{10}{260}\right)^{0.5}$$
$$= 98\,000\,\text{m}$$
$$= 98\,\text{km}$$

The resulting geostrophic wind can now be calculated with Equation (14.7) using $\rho_{bot} = 1.3\,\text{kg m}^{-3}$ and $\Delta p = 491\,$Pa from the previous example:

$$v_g = \frac{1}{1.3 \times (-1.4 \times 10^{-4})} \frac{-491}{98\,000}$$
$$= 28\,\text{m s}^{-1}$$

where the sign of the pressure gradient is negative since the pressure decreases toward the east, and the resulting geostrophic wind is from the south (positive v_g).

While this example is a rather extreme case, the blocking of a stable low-level flow can result in strong barrier parallel flows that are known as barrier winds.

Barrier winds often develop on the Ross Ice Shelf when a cyclone passes north of the ice shelf. The clockwise flow around the cyclone results in easterly winds across the Ross Ice Shelf. If this easterly flow is sufficiently weak and/or the stability is sufficiently large, Fr will be less than unity and the low-level flow will be blocked. An example of a barrier wind on the Ross Ice Shelf that developed in this manner on 11 April 2005 is shown in Figure 14.5. Similar situations give rise to barrier winds along the east side of the Antarctic Peninsula and along the steep margins of the Greenland ice sheet. These strong winds can create difficult conditions for travel and may influence the distribution of sea ice if they occur over the ocean.

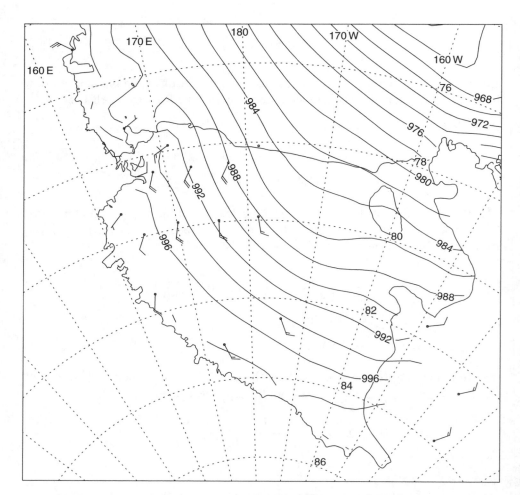

Figure 14.5 Surface weather map over Ross Ice Shelf, Antartica, for 12 UTC 11 Mar 2005. Sea level pressure contours are shown for locations with elevations less than 500 m and wind observations from automatic weather stations are plotted using the standard convention. Map provided by M. Seefeldt, CIRES/ATOC, University of Colorado

14.3 Polar lows

The category of mesoscale cyclones spans a broad range of weather systems, and one of the most intense types is the *polar lows*. These systems can be extremely intense, with winds comparable to those in a tropical cyclone and heavy snowfall. Like tropical cyclones, polar lows are relatively axisymmetric and warm cored, at least in the lower troposphere. Early interest in polar lows developed in Scandinavia, where small violent storms were known to affect both ships and coastal communities over very limited areas. Such storms were well known but impossible to predict. With the advent of the satellite era, these small, intense, damaging storms could finally be identified, tracked, and, to an extent, predicted. They appear to be a phenomenon of the eastern Arctic, around Greenland and the north Atlantic, being exceedingly rare on the Pacific side. Even more recently, such cyclones have been identified in the Southern Ocean around Antarctica, although they tend to be weaker due to the strong statically stable conditions.

An accepted definition of a polar low is that it is a maritime mesoscale cyclone with near surface winds in excess of $15\,\mathrm{m\,s^{-1}}$ that forms poleward of the polar front. Many methods have been used to classify the broad variety of systems that satisfy this definition, which includes systems that form in highly baroclinic regions along the polar front and near ice edges, as well as systems that form convectively in high-latitude, nearly barotropic environments. It is most accurate, therefore, to consider polar lows to be a spectrum of mesoscale systems, ranging from purely baroclinic to purely convectively driven, with a range of hybrid systems in between. Further, a particular polar low may owe its development to baroclinic instabilities or thermal instabilities at different stages of its life cycle. Additional details on polar lows can be found in Rasmussen and Turner (2003).

Example In February 1990, a meso-beta scale polar low passed close to the meteorological station located in Tromso, Norway (69°40′N, 18°56′E), which registered winds of $15.8\,\mathrm{m\,s^{-1}}$ at a time when atmospheric pressure was 1001 hPa. Estimate the central pressure of the cyclone.

Since the cyclone is assumed to be mesoscale and relatively axisymmetric, the gradient wind model is appropriate. We can assume an order of magnitude for the diameter of the cyclone to be 100 km. Then the model (Equation (6.1)) predicts

$$\frac{\partial p_d}{\partial n} = -\rho \left(\frac{V^2}{R} + fV\right)$$

$$= -1.2 \left(\frac{15.8^2}{50 \times 10^3} + 2 \times 7.292 \times 10^{-5} \times \sin 69.67° \times 15.8\right)$$

$$= -1.2 \left(4.99 \times 10^{-3} + 2.16 \times 10^{-3}\right)$$

$$= -8.58 \times 10^{-3}$$

$$p_d = -8.58 \times 10^{-3} \times 50 \times 10^3 + 100\,100$$
$$= 997\,\text{hPa}$$

14.3.1 Baroclinic instability

It has been widely accepted for some time that mid-latitude synoptic scale systems form through the process of baroclinic instability as described in Chapter 9. Detailed observations have made it possible to identify the various stages in the life cycle of a baroclinic development as expected by the theoretical model (Figure 1.9). Although in this model the lower level cyclone and upper level trough develop simultaneously as a continuous process, in fact initiation will happen at one level in the presence of a favorable environment in the other level. So, for example, a well-defined upper level short-wave trough may move over a pre-existing lower level, rather shallow, frontal zone. Alternatively, a fairly zonal upper level flow may encounter a particularly strong region of lower level baroclinicity, which supplies the energy to perturb the upper level flow. In some cases, polar lows develop in regions of *reverse shear*, where the horizontal wind decreases with height.

Consider the shallow baroclinic type. In this case, a shallow frontal zone may persist between a region of snow-covered land or sea ice and the relatively warmer open ocean. As an upper level trough passes over this frontal zone toward the warmer surface, the strong static stability of the polar atmosphere is lowered, perhaps even becoming neutral. In such conditions, cyclogenesis may take place. The lower the static stability, the smaller the horizontal scale of the system. In addition, the planetary vorticity is large at high latitudes, and this can contribute to rapid cyclogenesis.

In deep baroclinic zones, the development is more typical of mid-latitude cyclogenesis. In this process, a small initial perturbation in an otherwise uniform zonal flow will engender a wave-like structure in the zonal oriented isotherms. As we saw in Chapter 9, this wave is displaced a quarter wavelength to the west of the wave in pressure. Horizontal temperature advection due to the pressure perturbation and associated geostrophic wind then amplifies the temperature perturbation. This allows the development of a thermally direct circulation, in which cold air is sinking and warm air is rising, which converts available potential energy in the horizontal temperature gradient to kinetic energy of cyclogenesis.

14.3.2 Convection

It has been recognized since the earliest studies of polar lows that convection often plays a significant role, although the exact nature of this role has been disputed and remains an area of active research. Nevertheless, a fairly consistent picture has now emerged. The polar atmosphere, like the tropical one, is nearly neutral to deep moist convection. Hence, CAPE must be generated in order for a development to take place. This can be achieved by strong surface fluxes of latent and sensible heat that destabilize lower layers. Alternatively, or in addition to surface forcing, CAPE can be generated through the advection of cold air aloft. Hence, polar lows do not develop from existing reservoirs of CAPE, but rather co-develop with the CAPE reservoir as the air flows over a warm sea surface. The maximum observed CAPE value in a polar low has been observed to be around $1100\,\text{J}\,\text{kg}^{-1}$, and values around 400–$600\,\text{J}\,\text{kg}^{-1}$ are more typical. Such moderate values compared to mid-latitude thunderstorms are nevertheless a significant source of energy in most polar low developments. If CAPE is not consumed as rapidly as it is generated, the reservoir may build up and continue to feed the cyclogenesis over time.

As with a tropical cyclone, heating in the middle troposphere is the most effective way to enhance cyclone development, since it leads directly to a decrease in surface pressure. The formation of a warm core may arise from the surface turbulent heat transport noted above (the WISHE model, Section 12.3), a mechanism which requires no CAPE in the atmospheric profile. Alternatively, it has been suggested that polar low intensification may occur more explosively through the heat release associated with mid-tropospheric cumulus convection, which does require a reservoir of CAPE in the atmospheric profile. This conceptual model is called CISK (Conditional Instability of the Second Kind), and was once thought to be of primary importance in the development of tropical cyclones. It is likely that both instability processes contribute to polar low development in different cases and even at different stages of a single low.

It has been found that disturbances must be of a finite amplitude in order to intensify a polar low through the thermal instability mechanisms described above. Hence, it appears that the development of polar lows is a two-stage process. Initial development, then, occurs through a process of baroclinic instability, or arises from another disturbance such as a topographically generated cyclone. Then, thermal instability can play a role in the intensification of the disturbance into an intense polar low. This two-stage process is supported by satellite observations, particularly in the Arctic, and has led some authors to characterize thermal instabilities as a cyclone 'afterburner'. However, because upper level troughs are generally associated with cold air advection, which enables the generation of CAPE, the two processes may be thought of as cooperating to bring about the polar low.

Review questions

14.1 Using Equations (14.2) derive Equations (14.3). (Hint: to derive the two expressions for V use v times Equation (14.2a) minus u times Equation (14.2b) to derive one of the equations for V and u times Equation (14.2a) plus v times Equation (14.2b) to derive the other equation for V. The equation for $\cos\beta$ can be derived using the two expressions for V and by noting that $\sin^2\beta + \cos^2\beta = 1$.)

14.2 An idealized representation of the Antarctic ice sheet topography can be given by

$$z = 4000\,\text{m} \times \left(1 - \frac{1000\,\text{km} - x}{1000\,\text{km}}\right)^{0.5}$$

for 0 km $< x \leq$ 1000 km.

(a) Write an equation for the terrain slope between $x = 0$ and 1000 km.

(b) Plot the wind speed and deviation angle as a function of x for inversion strengths of 5, 10, and 20 K. Assume that $C_D = 1.67 \times 10^{-5}$ m^{-1}, $f = 1.4 \times 10^{-4}$ s^{-1}, and $\bar{\theta} = 250$ K.

14.3 Based on your answers to question 14.2(b) plot the ratio of the katabatic force to (a) the Coriolis force and (b) the frictional force as a function of x.

14.4 Based on your answers to question 14.2(b) plot the ratio of the downslope advection term ($= u\, \partial u/\partial x$) to the katabatic force term as a function of x. When deriving the governing equations for katabatic flow, is the assumption that the advection term is negligible reasonable?

14.5 Using the idealized two-layer model of katabatic winds, and assuming that there is no background pressure gradient force, indicate the expected wind direction, relative to an ice sheet topography, in the Northern Hemisphere for

(a) a balance between the katabatic and Coriolis forces;

(b) a balance between the katabatic and frictional forces;

(c) a three-way balance between the katabatic, Coriolis, and frictional forces.

14.6 Minna Bluff is an east/west-oriented ridge that extends from the Transantarctic Mountains onto the Ross Ice Shelf at a latitude of 78.5°S, with a height of 900 m. This ridge is often in the path of the climatological southerly winds that blow across the western side of the Ross Ice Shelf. Seefeldt et al. (2003) analyzed a case of southerly flow impinging on Minna Bluff on 22 April 1994. For this case, there were south winds at 5 m s^{-1}, the potential temperature difference between the surface of the Ross Ice Shelf and the top of Minna Bluff was 10 K, and the surface temperature over the Ross Ice Shelf was 248 K.

(a) Calculate the Froude number for this flow impinging on Minna Bluff.

(b) Assume that the depth of the undisturbed flow upstream from Minna Bluff is 500 m, and that the flow deepens to a depth of 750 m adjacent to Minna Bluff. What is the horizontal pressure difference between the undisturbed flow and the flow at the base of Minna Bluff?

(c) What is the horizontal distance over which the pressure change in part (b) occurs?

(d) Estimate the geostrophic wind speed and direction upstream from Minna Bluff for this case.

15 Epilogue: the general circulation

15.1 Fueled by the Sun

In this book we have been studying many aspects of the motion of the atmosphere, but we have not considered the ultimate source of energy for that motion. In fact, the atmosphere and ocean circulation that we observe on the Earth is a system that is fueled by the Sun – more specifically, by the imbalance of energy input to the Earth system at the equator and the poles (Figure 15.1).

Energy from the Sun is received as *short-wave radiation*, light of wavelengths between 0.2 and 4 μm. This is equivalent to a blackbody radiating at 5776 K, but with rather less ultraviolet than would be expected. About 40% of the incoming energy is in the visible part of the spectrum (0.4–0.67 μm) and about 10% is in the ultraviolet (< 0.4 μm). The energy flux emitted by the Sun is not perfectly constant and varies at a rate and over a range that is important for studies of climate, though not for meteorology and numerical weather prediction.

Even considering only time scales where the energy flux emitted by the Sun may be considered constant, the energy received at the top of the atmosphere varies with season and with latitude. The elliptical orbit of the Earth causes a variation of ±3.5%, and the tilt of the earth's axis, at an angle of 23.5° from the orbital plane results in seasonal variations which far exceed the variation arising from the elliptical orbit. When account is taken of these variations, the average flux in 1 year varies with latitude as shown in Figure 15.1.

All of the energy impinging on the Earth is not absorbed: a percentage is reflected or scattered. This percentage has an annual average value of about 30%, which consists of 6% back-scattered by the air, 20% reflected by clouds, and 4% reflected by the surface (land and oceans) itself. This fraction of reflected and scattered incoming solar radiation is known as the *albedo* (α). The albedo can vary enormously with space and time, because it depends on such factors as surface type and cloud distribution. Hence, the total short-wave energy absorbed by the system is not as symmetrical about the equator as the total incoming short-wave energy (Figure 15.1).

In the long term, the Earth system should neither gain nor lose heat. Hence all of the absorbed solar radiation is re-emitted back to space in the *long-wave* portion of

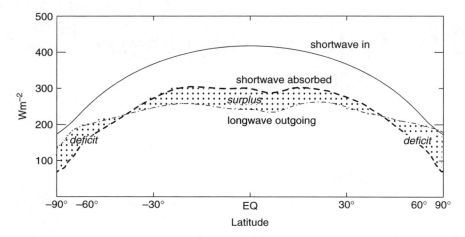

Figure 15.1 Average radiation balance for the Earth, showing incoming short-wave radiation (solid line), absorbed short-wave radiation (dashed line), and outgoing long-wave radiation (dot dash line). The horizontal axis shows the sine of the latitude, and is hence proportional to surface area. Stippled zones indicate latitudes of mismatch between annual mean energy in and out. For long-term balance areas of surplus must exactly balance areas of deficit. NCEP Reanalysis data provided by the NOAA-CIRES Climate Diagnostics Center, Boulder, Colorado, from its Web site at http://www.cdc.noaa.gov/

the spectrum (greater than around 4 μm). Of the total outgoing long-wave radiation, 37% is emitted by clouds, 54% is emitted by gases in the atmosphere such as water vapor and carbon dioxide (and much of this is long-wave energy that was absorbed in the atmosphere from the surface below), and 9% is emitted directly from the surface of the Earth, including the oceans. The important role of the atmosphere in the energy balance of the Earth system is apparent.

As for the absorbed solar energy, the emitted long-wave energy varies dramatically in space and time, creating a surplus of energy at some locations and times and a deficit at others. While this varies according to season and surface and with transient effects of features such as clouds and snow cover, the most systematic pattern of surplus and deficit is the variation with latitude. It is this energy imbalance which drives atmospheric and oceanic movement to transport energy; in Sections 15.2 and 15.3 we will look at some ways in which this imbalance can be corrected.

15.1.1 The greenhouse effect

Because of the absorption and re-emission of long-wave energy by the atmosphere, the average temperature at the surface of the Earth is around 33 K warmer than would be expected if the Earth had no atmosphere. This is because, while about 74% of short-wave energy entering the system passes through the atmosphere to the surface of the Earth, around 62% of long-wave energy emitted from the surface is reabsorbed in the atmosphere before it leaves the system. This energy is re-emitted of course, since balance must be maintained, but in so doing some energy is re-emitted downward to the surface of the Earth, and hence increases the temperature near the surface. This so-called 'greenhouse effect' creates the benign conditions for life on Earth that otherwise would require an orbit closer to the Sun.

The *enhanced greenhouse effect* is the process whereby long-wave radiation absorbers such as carbon dioxide and methane are added to the atmosphere in the course of human activities. In such a situation, the energy balance of the Earth system is modified, and with this comes changes in the circulation of the atmosphere and the ocean.

15.2 Radiative–convective equilibrium

The atmosphere is, in large part, heated from below due to the emitted long-wave radiation from the surface. Because of this source of heat from below, the resulting atmospheric profile must tend to be statically unstable and hence convective motions are generated. This aspect of the energy transformations in the Earth system led to the idea in the seventeenth century of a radiative–convective equilibrium, whereby air heated at the surface in the tropics rises convectively and flows to the poles, where it sinks and returns at surface level to the tropics. This results in a net transport of heat from equator to pole. The surface flow was identified as the *trade winds*, so called due to their consistency, which was good for rapid, reliable transport of goods by sailing ship. This conceptual model was first postulated in 1686 by English astronomer Edmund Halley. Halley suggested that the trade winds north of the equator blow from east to west due to a build-up of flowing air that follows the heat of the Sun as it travels from east to west due to the rotation of the Earth.

In 1735, professional lawyer and amateur meteorologist George Hadley made an important refinement to Halley's theory. Hadley correctly surmised that at the Earth's surface, the planet and the atmosphere rotate together. Because every latitude circle on the Earth completes one revolution in one day, this means that the air and the surface at higher latitudes rotate more slowly than air and the surface at the equator. So, wind moving toward the equator would come from a region of lower eastward velocity and hence result in a westward turning of the wind. Hadley's conclusions were published 100 years before Coriolis systematically described motions in a rotating coordinate system (Section 4.4.2). While neither Halley nor Hadley had all of the conceptual tools necessary to understand the dynamics of the trade winds, the role of the atmosphere in carrying excess heat to the poles was an important first step in understanding the global circulation.

Such a process, then, causes a simple *direct circulation* carrying heat from equator to pole (Figure 15.2), correcting the energy imbalance caused by the differential in short-wave absorption and long-wave emission. A direct circulation is one in which hot air rises and cool air sinks. Hadley's single-cell theory goes some way to explaining the behavior of the trade winds and tropical convection. In his conceptual model, the upward motion of the Hadley cell is driven by latent heat release as water vapor is converted into precipitation. The rising motion, then, is embedded within thunderstorms, which in fact occupy only around 0.5% of the tropical surface

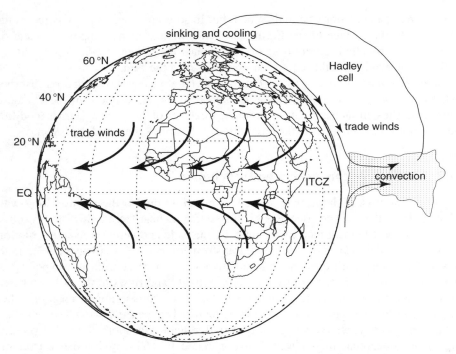

Figure 15.2 Hadley's postulated single-cell circulation which accounts for the trade winds and convection in the tropics. The zone of convection, called the Intertropical Convergence Zone (ITCZ), follows the latitude of maximum solar heating throughout the year. The cell illustrated is not to scale and only the Northern Hemisphere complete circulation is shown

area. However, models representing the single-cell theory predict an equator-to-pole temperature gradient of around 120 K, instead of the observed value closer to 45 K.

A further problem with the Halley/Hadley conceptual model is that it is not dynamically realistic on a rotating Earth. Conservation of angular momentum of the Earth–atmosphere system requires that winds from the same direction cannot exist at all locations on the Earth. Easterly winds in one location must be balanced by westerly winds somewhere else. This conservation requirement is discussed in further detail in Section 15.4.

15.3 The zonal mean circulation

The baroclinically driven mid-latitude cyclones explored in Part I cannot be accounted for in the Hadley single-cell model. Yet these systems derive their energy from the potential energy available in the mean meridional temperature gradient, and hence act to reduce this gradient. This implies that the observed mean temperature field is due to a balance between the competing effects of a differential radiation distribution and a broader range of instabilities.

An early attempt to address this problem was made by American meteorologist and oceanographer William Ferrel, who is best known for his work on tides. In 1856, he proposed a mid-latitude cell characterized by an *indirect circulation*, in which air at lower latitudes sinks and flows poleward and eastward, while cooler air at higher latitudes rises and flows equatorward and westward. This model does not match up precisely with the observations, and he supplied no mathematical justification. Nevertheless, like Halley before him, it was an important attempt to explain aspects of the global circulation (in this case, the mid-latitude westerlies) that laid the groundwork for future research. On this important foundation, new insights continued to be generated. For example, in 1921, German meteorologist and oceanographer Albert Defant suggested that traveling mid-latitude cyclones and anticyclones could be viewed as turbulent elements in a roughly horizontal process of heat exchange between air masses. This idea turned out to be pivotal in analyses of the angular momentum and energy budgets of the Earth. Then, in the 1930s, Carl Gustav Rossby relied on the work of Hadley, Ferrel, and Defant, among others, to create the more complete conceptual model shown in Figure 15.3.

The strength of the Hadley circulation in this model varies with longitude, being strongly affected by topography and land/sea contrast. The rising motion at the ITCZ

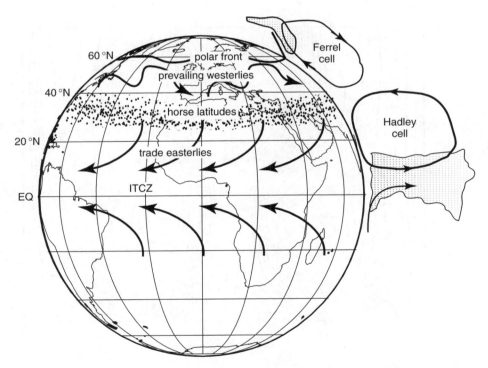

Figure 15.3 A revised picture of the zonal mean circulation, based on Rossby's work of the 1930s and 1940s

is a region also called the *doldrums*, because prevailing horizontal winds are light and variable. After rising near the equator, the air in the Hadley cell moves poleward in the upper region of the troposphere, sinking near 30° (N or S) and generating belts of high pressure. The subtropical jet streams arise near the upper, poleward portions of the circulation. The circulation is completed by the equatorward flow of the trade winds. The downward branch of the Hadley cell is associated with the world's great deserts (e.g. Mojave Desert, the Gibson Desert, the Sahara Desert, the Gobi Desert, etc.).This region of light and variable winds at the downward branch is known as the *horse latitudes*, although there are various explanations as to the origin of this term.

In Rossby's original work, he suggested a second direct circulation at the poles, to create what was called the 'three-cell model'. However, the polar cell, if it can be defined, is rather weak (Figure 15.4a–c). The Ferrel cell should be understood as a simple Eulerian mean that does not represent the actual motion of air parcels. Nevertheless, without Rossby waves and the associated baroclinic developments, the Ferrel cell would not be observed in the zonal mean.

Example Use the principle of thermal wind balance to determine the expected zonal mean wind $\bar{u}$ at the Northern Hemisphere jet maximum in Figure 15.4(a). How does this compare to the jet maximum in a single cell, or Hadley, regime?

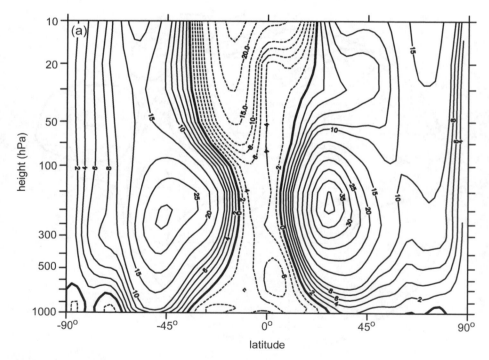

Figure 15.4(a) Aspects of the observed zonal mean general circulation: the zonal mean zonal wind (m s^{-1}) with westerlies shown as positive and easterlies shown as negative. This is a time mean of 20 vernal equinoxes (21 March). The jet maxima at around 200 hPa are apparent

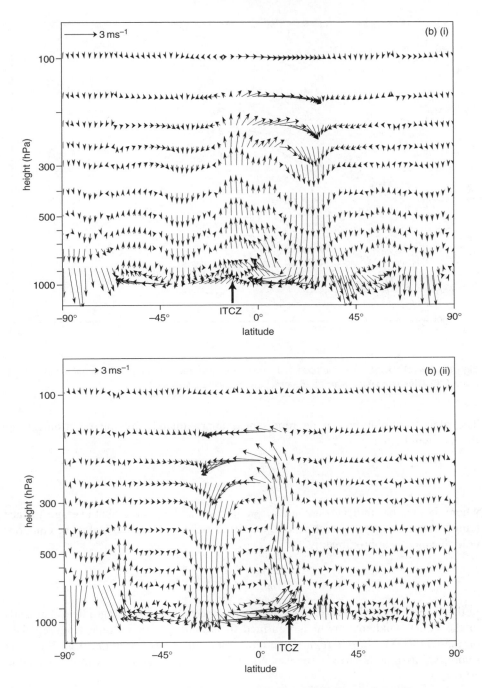

Figure 15.4(b) Aspects of the observed zonal mean general circulation: the zonal mean meridional circulation for (i) December–January–February and (ii) June–July–August averages. NCEP Reanalysis data provided by the NOAA-CIRES Climate Diagnostics Center, Boulder, Colorado, from its Web site at http://www.cdc.noaa.gov/

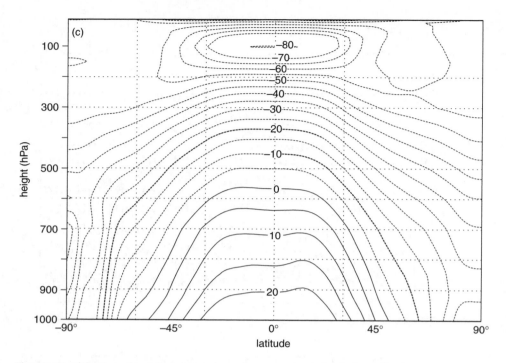

Figure 15.4(c) Aspects of the observed zonal mean general circulation: the zonal mean temperature (°C). This is a time mean of 20 vernal equinoxes (21 March)

The global circulation can be represented approximately by the requirement for thermal wind balance:

$$f_0 \frac{\partial \overline{u}}{\partial z} = -\frac{g}{T_0} \frac{\partial \overline{T}}{\partial y}$$

where the overbar indicates a zonal mean quantity and T_0 is a global average surface temperature. The mean radius, a, of the Earth is 6.37×10^6 m, and hence the equator–pole distance is approximately

$$\Delta y = \frac{2\pi a}{4} = 1.00 \times 10^7 \, \text{m}$$

In the case of the observed jet (see Figure 15.4a), the temperature gradient that leads to the zonal wind maximum is confined to around 20° of latitude (Figure 15.4d on the CD-ROM). In this region, the temperature decreases around 20°C. From the figure, we assume a surface zonal wind of $2 \, \text{m s}^{-1}$, and the jet maximum at around 10 km,

$$\frac{\Delta \overline{T}}{\Delta y} = \frac{-20}{0.222 \times 10^7} = -9.0 \times 10^{-6} \, \text{K m}^{-1}$$

$$\frac{\Delta \bar{u}}{\Delta z} = -\frac{g}{f_0 T_0} \frac{\Delta \bar{T}}{\Delta y}$$

$$= -\frac{9.81}{1.0 \times 10^{-4} \times 288} \times -9.0 \times 10^{-6}$$

$$= 3.1 \times 10^{-3}$$

$$\Delta \bar{u} = 31 \, \mathrm{m\,s^{-1}}$$

$$\bar{u}_{jet} = 33 \, \mathrm{m\,s^{-1}}$$

which accords quite well with the observed jet. In the case of the Hadley regime,

$$\frac{\Delta \bar{T}}{\Delta y} = \frac{-120}{1.00 \times 10^7} = -1.2 \times 10^{-5} \, \mathrm{K\,m^{-1}}$$

and hence the predicted jet will be around 30% stronger than observed. In fact, if the jet is similarly confined to a smaller latitudinal band, this jet may be somewhat stronger.

Rossby's work has been refined since then by scientists such as Victor Starr, Norman Phillips, and Edward Lorenz. Their important insights led to the current view that the Hadley circulation is driven primarily in fact through cooling by transient baroclinic waves in the mid-latitude storm tracks. It is for this reason that the Hadley cells translate with season. Warm tropical oceans do determine where upward motion is favored, but the primary source of moisture is transported by trade winds from the subtropics. Nevertheless, the fundamental view that the Hadley circulation is the most efficient means of poleward heat transport in the tropical regions remains.

The thermal wind balance requirement imposes a strong constraint on the zonal mean circulation, in that any pressure gradient that results from a departure from balance will drive a mean meridional circulation which adjusts the mean zonal wind and temperature fields so that thermal wind balance is restored. Thus, the mean meridional circulation plays the same role in the zonal mean circulation that the ageostrophic flow plays in synoptic scale quasi-geostrophic systems.

15.4 The angular momentum budget

Because there are no external sources of torque, the total angular momentum of the Earth–atmosphere system must be conserved. Since the average rate of rotation of the Earth is approximately constant, the atmosphere must also on average conserve its angular momentum. The role of the ocean in this angular momentum budget is becoming better understood at present.

The mid-latitude prevailing westerlies are an important part of the global circulation. These winds circulate around the Earth in the same sense as the Earth's rotation. Conversely, the tropical easterlies rotate less rapidly than the rotation of the Earth.

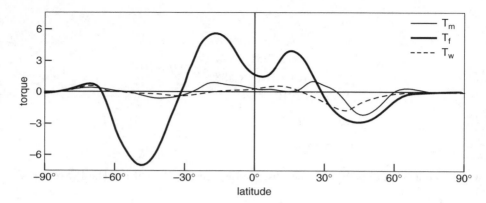

Figure 15.5 Annual mean zonally integrated surface pressure or mountain torque (Tm), surface friction torque (Tf), and gravity wave drag torque (Tw) in hadleys (10^{18} kg m^2 s^{-2}). With perfect data, these torques would be exactly balanced by the annual mean angular momentum transport. NCEP Reanalysis data provided by the NOAA-CIRES Climate Diagnostics Center, Boulder, Colorado, from its Web site at http://www.cdc.noaa.gov/

To maintain balance, angular momentum that is gained in the tropics due to the interaction between surface and atmosphere must be transported poleward and transferred back to the Earth at higher latitudes.

The transfer of angular momentum between the Earth and atmosphere is achieved through the action of three torques (which is the force multiplied by the distance from the axis of rotation). These torques are caused by (i) surface friction, (ii) the pressure differential between the leeward and windward sides of meridionally oriented mountain ranges, and (iii) vertical momentum transfer due to internal gravity waves (Figure 15.5). Until the late 1980s, the role of gravity wave stresses was not known, and the difference between torques and angular momentum transport was attributed to measurement error. In the middle latitudes of the Northern Hemisphere the surface pressure differential or 'mountain torque' provides nearly half of the total atmosphere surface momentum exchange. In the tropics and the Southern Hemisphere the exchange is dominated by surface friction and the turbulent eddy stresses it engenders.

This angular momentum budget is balanced by the poleward transport of angular momentum, which is achieved when poleward moving air carries more angular momentum than equatorward moving air. In the equatorial regions a large fraction of this transport is done by the zonal mean circulation: the Hadley cell. In the Hadley cell, parcels containing the angular momentum transferred from the surface are lifted to the upper troposphere and poleward, while parcels with less angular momentum sink and move equatorward, resulting in a net poleward transport.

Example Use the principle of conservation of angular momentum to determine the expected zonal mean wind $\bar{u}$ at the Northern Hemisphere jet maximum in Figure 15.4(a).

Angular momentum is given by ωr^2, where $\omega = \bar{u}/r$ is the total rate of rotation of the atmosphere relative to an observer in an inertial frame, and r is the distance to the axis of rotation. Hence,

$$\left(\Omega + \frac{\bar{u}_{30°N}}{r_{30°N}}\right) r^2_{30°N} = \left(\Omega + \frac{\bar{u}_{eq}}{r_{eq}}\right) r^2_{eq}$$

$$\bar{u}_{30°N} = r_{30°N} \left[\left(\Omega + \frac{\bar{u}_{eq}}{r_{eq}}\right) \frac{r^2_{eq}}{r^2_{30°N}} - \Omega\right]$$

$$= a \cos 30° \left[\left(7.292 \times 10^{-5} + \frac{-4}{a}\right) \frac{a^2}{a^2 \cos^2 30°}\right.$$

$$\left. -7.292 \times 10^{-5}\right]$$

As noted above, the radius of the Earth, a, is 6.37×10^6 m, and hence we calculate that $u_{30°N} = 129$ m s^{-1}. This is much larger than the observed jet maximum, and suggests that the zonal mean circulation alone cannot be responsible for the angular momentum transport. This reflects the importance of what is termed the *eddy momentum flux*: that is, the momentum transport by disturbances that may be described as transient and small scale relative to the scale of the zonal mean circulation.

Hence, weather systems, though transient and small scale compared to the global circulation, are an important component of angular momentum transport. In middle latitudes, they predominate. The mid-latitude eddy angular momentum transport is a maximum at upper levels (around the jet stream), and is accomplished through the pronounced eastward tilt with increasing latitude of the horizontal flow in weather systems (Figure 15.6). This important asymmetry means that air flowing poleward will have a stronger westerly wind component than the equatorward return flow.

15.5 The energy cycle

We have seen that the requirements for thermal wind balance and for angular momentum conservation give us just part of the story when we consider only the components of the large-scale circulation. Now that we turn our attention to the energy cycle, we will find once again the central role for synoptic weather systems in the functioning of the Earth's atmospheric dynamics.

Like angular momentum, energy is conserved and hence can be followed as it transforms from one form to another. The total energy of a parcel in the atmosphere is made up of the kinetic energy of its motion, arising predominantly from the horizontal component, and the *moist static energy* h, which is given by

$$h = c_p T + \Phi + Lr \qquad (15.1)$$

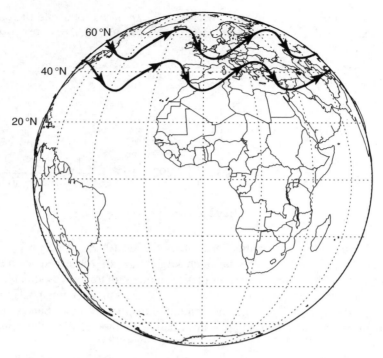

Figure 15.6 The relationship between the tilt of mid-latitude waves at jet stream level and the net poleward transport of positive angular momentum

where L is the latent heat of condensation, r is the mixing ratio, Φ is the geopotential, T is the temperature, and c_p is the specific heat at constant pressure of dry air. The first two terms on the right hand side of Equation (15.1) are generally grouped together as the total potential energy, or *dry static energy*, of the parcel. Around 0.1% of the total potential energy, termed the *available potential energy*, is accessible to be extracted and converted into kinetic energy. The available potential energy, in an important insight by meteorologist Edward Lorenz, arises from the equator-to-pole temperature gradient, and is the source of energy for synoptic disturbances, as first discussed in Section 9.3. Because the available potential energy resides in the temperature *gradient*, diabatic heating or cooling will only increase the available potential energy if it enhances that gradient. If the heating or cooling acts to decrease the temperature gradient, then available potential energy is actually decreased.

In the long-term mean, the production of available potential energy must exactly balance the dissipation through turbulent and thence frictional processes (via the energy cascade). However, this balance is not maintained in regional or temporal subsets, and can only be considered as a global, long-term budget. The derivation of the equations that describe this budget are beyond the scope of this book, but they may be summarized and illustrated in a 'box diagram', an approach pioneered by Lorenz in 1955 and shown in Figure 15.7.

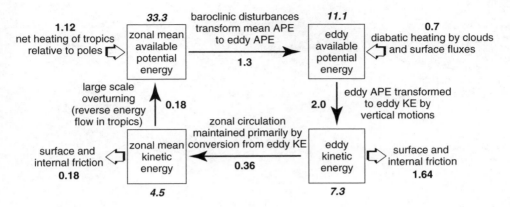

Figure 15.7 Lorenz energy cycle for the global atmosphere in the annual mean. Estimates of energy density in $10^5 \, \text{J m}^{-2}$ are given in italics at each box and rates of energy transfer (that is, generation, conversion, and dissipation) are given in W m^{-2} at each arrow. Quantities are estimates only, and based on the work of Abraham Oort and José Peixoto using a 10 year mean of meteorological station data

This diagram is based on station data, which limits the accuracy due to problems of missing data, large gaps over the oceans, and an inability to fully account for vertical motion. The direction of the flow of energy between reservoirs of available potential and kinetic energy must be deduced rather than observed. Even in more recent analyses using global gridded datasets, considerable differences between datasets and methods arise. However, these analyses all reflect the fact that radiative heating from the Sun and moist diabatic processes represent a net influx of energy into the system, which ultimately is balanced by frictional dissipation. Some versions of this analysis split the kinetic energy into stationary and transient components, since the effects of forced orographic waves and stationary planetary waves also contribute substantially to the poleward heat flux. In fact, it is apparent in recent studies that there is compensation and mutual adjustments between the stationary and transient components of energy transport, resulting in a total energy transport as shown in Figure 15.7. While all attempts to characterize the global energy cycle are limited by the many assumptions made in deriving them and, as always, by limited data, they do lend considerable insight into the processes maintaining the general circulation. In particular, the energy cycle highlights the primary role of weather disturbances.

Appendix A – symbols

a – average radius of the Earth (constant): 6.37×10^6 m
$\vec{a}$ – acceleration vector
A – area (also unit area, also wave amplitude, also coefficient or constant of integration to be solved for per boundary conditions in Ekman layer theory)
b – constant in Teten's formula: $= 17.27$ K^{-1}
B – coefficient (constant of integration) to be solved for per boundary conditions in, for example, Ekman layer theory
c – phase speed
C – circulation
$C_{absolute}$ – absolute circulation, including contributions from planetary rotation and from other sources of rotation relative to the surface of the planet
CAPE – Convective Available Potential Energy
C_D – drag coefficient
C_{earth} – circulation resulting from the Earth's rotation
CF_{total} – total centrifugal 'force'
c_{gx} – zonal group velocity
c_{gy} – meridional group velocity
c_{gz} – vertical group velocity
c_p – specific heat at constant pressure (constant): 1004 J kg^{-1} K^{-1} (for dry air); 1952 J kg^{-1} K^{-1} (for water vapor)
C_p – phase speed, or speed of propagation, of ocean waves
$C_{relative}$ – circulation due to sources of rotary motion relative to the surface of a planet
c_v – specific heat at constant volume (constant): 717 J kg^{-1} K^{-1} (for dry air); 1463 J kg^{-1} K^{-1} (for water vapor)
c_x – zonal phase speed
$\hat{c}_x$ – intrinsic zonal wave propagation speed
c_y – meridional phase speed
c_z – vertical phase speed

d – channel width
DC – wind directional constancy
d/dv – derivative with respect to variable v
D/Dv – Lagrangian, substantial, or material derivative with respect to variable v
$\partial/\partial v$ – local, Eulerian, or partial derivative with respect to variable v
D_g/Dv – geostrophic Lagrangian derivative with respect to variable v
D_h/Dv – horizontal Lagrangian derivative with respect to variable v
e – vapor pressure
EL – Equilibrium Level
e_s – saturation vapor pressure
e_{si} – saturation vapor pressure over ice
f – Coriolis parameter, defined as $2\Omega \sin \phi$
f_0 – 'mid-latitude' reference Coriolis parameter, defined as $2\Omega \sin 45°$
$\vec{F}$ – force vector
$F_{Coriolis}$ – Coriolis force
Fr – Froude number
$\vec{F}_r$ – viscous force vector
g – effective gravity (gravitational acceleration plus centrifugal effect) (constant): 9.81 m s^{-2} at sea level at equator
$\vec{g}^*$ – gravitation vector
g_0^* – average gravitational acceleration at sea level (constant): 9.83 m s^{-2}
G – gravitation constant: 6.673×10^{-11} N m^2 kg^{-2}
h – lower boundary height as a function of location x, y; frontal height as a function of distance x; constant in Ekman layer theory $h = \sqrt{2K_m/f}$; moist static energy
H – scale height $H = R_d \langle T \rangle /g^*$; high-pressure center; vertical length scale
h_m – barrier height
$\vec{i}$ – unit x direction vector in a Cartesian coordinate system
$\vec{i}'$ – unit x direction in rotated Cartesian coordinate system
I – moment of inertia
$\vec{j}$ – unit y direction vector in a Cartesian coordinate system
$\vec{j}'$ – unit y direction in rotated Cartesian coordinate system
$\vec{k}$ – unit z direction vector in a Cartesian coordinate system; wavenumber vector
$\vec{k}'$ – unit z direction in rotated Cartesian coordinate system
k – zonal wavenumber
K_m – eddy viscosity
$\vec{k}_s$ – stationary wavenumber
L – length scale; low-pressure center
l – meridional wavenumber; ocean wave length
l^2 – Scorer parameter
LCL – Lifting Condensation Level
LFC – Level of Free Convection
m – mass; vertical wavenumber
M – mass of the Earth (constant): 5.9742×10^{24} kg; molecular mass

M_{air} – mass of a volume of air
M_d – average molecular mass for dry air (constant): $28.966\,\text{g}\,\text{mol}^{-1}$
M_v – average molecular mass for water vapor (constant): $18.016\,\text{g}\,\text{mol}^{-1}$
n – normally oriented coordinate in a natural coordinate system
N – Brunt–Väisälä or buoyancy frequency
p – pressure
p_0 – surface pressure; reference pressure
P – Ertel's potential vorticity
p_d – perturbation, or dynamic, pressure
$\vec{P}_g$ – pressure gradient force vector
p_n – pressure of the nth gas in a mixture
p_z – pressure at height z
q – specific humidity
Q – heat energy
r – mixing ratio; radius of curvature
R – gas constant for a particular gas; radius of curvature
R^* – universal gas constant: $8.314\,\text{J}\,\text{mol}^{-1}\,\text{K}^{-1}$
R_d – gas constant for dry air: $R^*/M_d = 287\,\text{J}\,\text{kg}^{-1}\,\text{K}^{-1}$
Re – Reynolds number
RH – Relative Humidity
R_n – gas constant for the nth gas in a mixture
Ro – Rossby number
r_R – Rossby radius of deformation
r_s – saturation mixing ratio
R_v – gas constant for water vapor: $461\,\text{J}\,\text{kg}^{-1}\,\text{K}^{-1}$
R_ϕ – radial distance from rotation axis to surface of the Earth at a given latitude
s – tangentially oriented coordinate in a natural coordinate system
$\vec{S}$ – vector describing a smooth curve
t – time
T – temperature; time scale; period of oscillation
T_0 – reference temperature, most commonly a function of z
T_{00} – constant reference temperature
T_1 – constant in Teten's formula: $273.16\,\text{K}$
T_2 – constant in Teten's formula: $35.86\,\text{K}$
T_A – temperature of parcel A
T_d – dew point temperature
T_{env} – temperature of the environment
T_{parcel} – temperature of a parcel
$T_{parcel,z}$ – temperature of a parcel at height z
T_v – virtual temperature
T_w – wet-bulb temperature
T_z – temperature at height z
u – zonal wind component
u_0 – reference zonal velocity component

U – horizontal velocity scale
$\vec{u}$ – velocity vector
$\hat{u}$ – zonal velocity amplitude
$\hat{u}_0$ – equatorial zonal velocity amplitude
$\vec{u}'$ – velocity vector in a rotated coordinate system
$\vec{u}_a$ – ageostrophic wind vector
u_a – zonal ageostrophic wind component
$\vec{u}_{earth}$ – velocity vector attributable to motion due to rotation of the Earth
u_i – velocity at a reference level i
$\vec{u}_g$ – geostrophic velocity vector
$\vec{u}_h$ – horizontal velocity vector
u_x – x-directed component of the velocity vector
v – meridional wind component
V – wind speed in natural coordinates; volume
$\vec{V}$ – wind vector in a natural coordinate system $= V\vec{\tau}$
v_a – meridional ageostrophic wind component
V_g – geostrophic wind in a natural coordinate system
v_i – velocity at a reference level i
w – vertical wind component
w_0 – amplitude of vertical velocity perturbation
W – vertical velocity scale
WD – wind direction
WS – wind speed
x – $\vec{i}$-directed Cartesian coordinate value
x_0 – reference $\vec{i}$-directed Cartesian coordinate value
y – $\vec{j}$-directed Cartesian coordinate value
y_0 – reference $\vec{j}$-directed Cartesian coordinate value
z – $\vec{k}$-directed Cartesian coordinate value; height; vertical coordinate in a natural coordinate system
Z – geopotential height
z_0 – reference $\vec{k}$-directed Cartesian coordinate value
α – angle; specific volume; phase constant for simple harmonic motion
β – meridional gradient of Coriolis parameter $\partial f/\partial y$; ocean wave phase speed constant of proportionality
δ – perturbation or variation
Δ – change or difference
$\delta\psi$ – angle subtended through a curve δs, the path taken in the tangential direction in the natural coordinate system
ε – frontal slope in Margules formula; also ratio of molecular mass of dry air to molecular mass of water vapor: 0.622
ζ – vorticity
ζ_{earth} – vorticity due to the Earth's rotation, f
ζ_g – geostrophic vorticity

ζ_{g0} – geostrophic vorticity at the surface
ζ_θ – vorticity on a constant potential temperature surface, isentropic vorticity
$\vec{\eta}$ – normal unit vector in a natural coordinate system
θ – potential temperature; angle of rotation of a pendulum
θ_0 – amplitude of angular rotation of pendulum
$\theta_{parcel,z}$ – potential temperature of a parcel at height z
θ_w – wet-bulb potential temperature
θ_z – potential temperature of air at height z
λ – wavelength
μ – dynamic viscosity
ν – kinematic viscosity: $1.5 \times 10^{-5}\,\mathrm{m^2\,s^{-1}}$ (Earth atmosphere), $1.0 \times 10^{-6}\,\mathrm{m^2\,s^{-1}}$ (Earth ocean); frequency
$\hat{\nu}$ – intrinsic frequency of oscillation
π – pi (constant): 3.141 59
ρ – density
ρ_0 – surface density; reference density, most commonly a function of z
ρ_{00} – constant reference density
ρ_i – density at a reference level i
ρ_n – density of nth gas in a mixture
$\rho_{parcel,z}$ – density of a parcel at height z
ρ_v – density of water vapor contained in a volume of air
ρ_z – density at height z
σ – buoyancy force
$\vec{\tau}$ – stress vector; tangentially directed axis in natural coordinate system
τ_x – x-directed stress vector component
τ_y – y-directed stress vector component
v' – perturbed v, where v is any variable or operator
$\bar{v}$ – averaged v, where v is any variable or operator
v_L – variable or parameter in a defined lower layer in a two-layer model
v_U – variable or parameter in a defined upper layer in a two-layer model
ϕ – latitude; smooth differentiable scalar function; phase of a simple harmonic oscillator
Φ – geopotential
ψ – deviation angle of the wind from the terrain fall line
ω – angular velocity
Ω – angular velocity of the Earth's rotation (constant): $7.292 \times 10^{-5}\,\mathrm{s^{-1}}$

Appendix B – constants and units

Useful constants
a – average radius of the Earth: 6.37×10^6 m
c_p – specific heat at constant pressure: $1004 \, \text{J kg}^{-1} \, \text{K}^{-1}$ (for dry air); $1952 \, \text{J kg}^{-1} \, \text{K}^{-1}$ (for water vapor)
c_v – specific heat at constant volume: $717 \, \text{J kg}^{-1} \, \text{K}^{-1}$ (for dry air); $1463 \, \text{J kg}^{-1} \, \text{K}^{-1}$ (for water vapor)
g – effective gravity (gravitational acceleration plus centrifugal effect): $9.81 \, \text{m s}^{-2}$ at sea level at equator
g_0^* – average gravitational acceleration at sea level: $9.83 \, \text{m s}^{-2}$
G – gravitation constant: $6.673 \times 10^{-11} \, \text{N m}^2 \, \text{kg}^{-2}$
M – mass of the Earth: 5.9742×10^{24} kg
M_d – average molecular mass for dry air: $28.966 \, \text{g mol}^{-1}$
M_v – average molecular mass for water vapor: $18.016 \, \text{g mol}^{-1}$
R^* – universal gas constant: $8.314 \, \text{J mol}^{-1} \, \text{K}^{-1}$
R_d – gas constant for dry air: $R^*/M_d = 287 \, \text{J kg}^{-1} \, \text{K}^{-1}$
R_v – gas constant for water vapor: $R^*/M_v = 461 \, \text{J kg}^{-1} \, \text{K}^{-1}$
ν – kinematic viscosity: $1.5 \times 10^{-5} \, \text{m}^2 \, \text{s}^{-1}$ (Earth atmosphere); $1.0 \times 10^{-6} \, \text{m}^2 \, \text{s}^{-1}$ (Earth ocean)
π – pi: 3.141 59
Ω – angular velocity of the Earth's rotation: $7.292 \times 10^{-5} \, \text{s}^{-1}$

Units
Primary SI units (Système International d'Unités)
Length: meter (m)
Mass: kilogram (kg)

Temperature: kelvin (K)
Time: second (s)

Derived units
Force: newton (N): $= \text{kg m s}^{-2}$
Pressure: pascal (Pa): $= \text{N m}^{-2} = \text{kg m}^{-1}\text{s}^{-2}$
Energy: joule (J): $= \text{kg m}^2 \text{s}^{-2}$
Power: watt (W): $= \text{kg m}^2 \text{s}^{-3}$

Other commonly used units in atmospheric science
Pressure: millibar (mb): $1\,\text{mb} = 100\,\text{Pa} = 1\,\text{hPa}$
Pressure: hectopascal (hPa): $1\,\text{hPa} = 100\,\text{Pa} = 1\,\text{mb}$
Temperature interval: degree Celsius (°C) : $1\,°\text{C} = 1\,\text{K}$
Temperature: Celsius degrees (°C) : $°\text{C} = \text{K} - 273.15$
Temperature: Celsius degrees (°C) : $°\text{C} = (°\text{F} - 32°\text{F})/(1.8\,°\text{F}\,°\text{C}^{-1})$
Temperature: Fahrenheit degrees (°F) : $°\text{F} = [°\text{C} \times (1.8\,°\text{F}\,°\text{C}^{-1})] + 32\,°\text{F}$
Wind Speed: knots (kts): $1\,\text{kt} = 0.51\,\text{m s}^{-1}$
Wind Speed: miles per hour (mph): $1\,\text{mph} = 0.48\,\text{m s}^{-1}$
Wind speed: kilometers per hour: (kph): $1\,\text{kph} = 0.28\,\text{m s}^{-1}$

Bibliography

Anderson, Jr. J. D., 1997: *A History of Aerodynamics: And Its Impact on Flying Machines.* Cambridge University Press, Cambridge.

Angel, W. (Ed.), 2003: Storm Data February 2003, National Oceanic and Atmospheric Administration, Asheville, NC.

Baines, P. G., 1995: *Topographic effects in stratified flows.* Cambridge University Press, Cambridge.

Ball, F. K., 1960: Winds on the ice slopes of Antarctica. *Antarctic Meteorology, Proceedings of the Symposium*, Melbourne, 1959, Pergamon, Oxford, pp. 9–16.

Batchelor, G. K., 1967: *An Introduction to Fluid Dynamics*, Cambridge University Press, Cambridge.

Bergeron, T., 1928: On linked three dimensional weather analysis. *Geofys. Publ.*, **5**, (6).

Bjerknes, J. and H. Solberg, 1922: Life cycle of cyclones and the polar front theory of atmospheric circulation. *Geofys. Publ.*, **3**, 3–18.

Blumen, W. and J. K. Lundquist, 2001: Spin-up and spin-down in rotating fluid exhibiting inertial oscillations and frontogenesis, *Dyn. Atmos. Oceans*, **33**, 219–237.

Brunt, D. and C. K. M. Douglas, 1928: The modification of the strophic balance for changing pressure distribution, and its effect on rainfall. *Mem. R. Meteorol. Soc.*, **3**, 29–51.

Charney, J. G. and A. Eliassen, 1949: A numerical method for predicting the perturbations of the middle latitude westerlies. *Tellus*, **1**, 38–54.

Cordero, E. C. and T. R. Nathan, 2000: The influence of wave- and zonal mean-ozone feedbacks on the quasi-biennial oscillation. *J. Atmos. Sci.*, **57**, 3426–3442.

Durran, D. R., 1990: Mountain waves and downslope windstorms. *Atmospheric processes over complex terrain*, W. Blumen, Ed., *American Meteorological Society*, Boston, MA, pp. 59–81.

Gill, A. E., 1982: *Atmosphere-Ocean Dynamics*, Academic Press, San Diego, CA.

Grant, A. M., 1957: A corrected mixing-length theory of turbulent diffusion. *J. Atmos. Sci.*, **14**, 297–303.

Holland, G. J. (Ed.), 1993: Global Guide to Tropical Cyclone Forecasting, WMO/TC-No. 560, Report No. TCP-31, World Meteorological Organization, Geneva.

Holton, J. R., 2004: *An Introduction to Dynamic Meteorology*, 4th ed. Elsevier, Burlington, MA.

Hunkins, K., 1966: Ekman drift currents in the Arctic Ocean, *Deep-Sea Res.*, **13**, 607–620.

Lorenz, E. N., 1955: Available potential energy and the maintenance of the general circulation. *Tellus*, **7**, 157–167.

Lynch, P., 2003: Margules' tendency equation and Richardson's forecast. *Weather*, **58**, 186–193.

Manabe, S. and R. F. Strickler, 1964: Thermal equilibrium of the atmosphere with a convective adjustment. *J. Atmos. Sci.*, **21**, 361–385.

Margules, M., 1906: Über Temperaturschichtung in stationär bewegter und in ruhender Luft (On temperature stratification in steadily moving and calm air). *Hann-Band. Meteorol. Z.*, 243–254.

Parish, T. R. and D. H. Bromwich, 1987: The surface windfield over the Antarctic ice sheets. *Nature*, **328**, 51–54.

Persson, A., 1998: How do we understand the Coriolis force? *Bull. Am. Meteorol. Soc.*, **79**, 1373–1385.

Petterssen, S., 1940: *Weather Analysis and Forecasting: A Textbook on Synoptic Meteorology*. McGraw-Hill, New York.

Pielke, Jr. R. A. and C. W. Landsea, 1998: Normalized Atlantic hurricane damage 1925-1995. *Weather Forecasting*, **13**, 621–631.

Prandtl, L., 1905: Über Flüssigkeitsbewegungen bei sehr kleiner Reibung (On fluid flow with very low viscosity). Verhandlungen des 3. internationalen Mathematiker-Kongresses in Heidelberg vom 8. bis 13. August 1904. Ed. A. Krazer. Teubner, Leipzig, pp. 484–491. (Repr. in Prandtl (1961) **2**: 575–584.)

Prandtl, L., 1961: Gesammelte Abhandlungen zur angewandten Mechanik, Hydro- und Aerodynamik (Collected papers in applied mechanics, hydraulics and aerodynamics). 3 vols. Springer, Berlin.

Rasmussen, E. A. and J. Turner, 2003: *Polar Lows: Mesoscale Weather Systems in the Polar Regions*. Cambridge University Press, Cambridge.

Richardson, L. F., 1922: *Weather Prediction by Numerical Process*. Cambridge University Press, Cambridge. (Repr. by Dover Publications, New York, 1965.)

Rogers, R. R. and M. K. Yau, 1989: *A Short Course in Cloud Physics*, 3rd ed. Butterworth–Heinemann, Oxford.

Salby, M. L., 1996: *Fundamentals of Atmospheric Physics*. Academic Press, San Diego, CA.

Schultz, D. M. and C. F. Mass, 1993: The occlusion process in a mid-latitude cyclone over land. *Mon. Weather Rev.*, **121**, 918–940.

Schultz, D. M., D. Keyser, and L. F. Bosart, 1998: The effect of large-scale flow on low-level frontal structure and evolution in midlatitude cyclones. *Mon. Weather Rev.*, **126**, 1767–1791.

Seefeldt, M. W., G. J. Tripoli, and C. R. Stearns, 2003: A high-resolution numerical simulation of the wind flow in the Ross Island region, Antarctica. *Mon. Weather Rev.*, **131**, 435–458.

Sheridan, S. C., 2002: The redevelopment of a weather-type classification scheme for North America. *Int. J. Climatol.*, **22**, 51–68.

Smagorinsky, J., 1963: General circulation experiments with the primitive equations. *Mon. Weather Rev.*, **91**, 99–164.

Smith, R. K., 1990: Translation of balanced air mass models of fronts and associated pressure changes. *Mon. Weather Rev.*, **118**, 1922–1926.

Smith, R. K. and M. J. Reeder, 1988: On the movement and low-level structure of cold fronts. *Mon. Weather Rev.*, **116**, 1927–1944.

Starr, V. P., 1948: On the production of kinetic energy in the atmosphere. *J. Meteorol.*, **5**, 193–196.

Stoelinga, M. T., J. D. Locatelli, and P. V. Hobbs, 2002: Warm occlusions, cold occlusions and forward-tilting cold fronts. *Bull. Am. Meteorol. Soc.*, **83**, 709–721.

Sutcliffe, R. C., 1938: On development in the field of barometric pressure. *Q. J. R. Meteorol. Soc.*, **64**, 495–504.

Wallace, J. M. and P. V. Hobbs, 2005: *Atmospheric Science: An introductory survey*, 2nd ed. Elsevier, Burlington, MA.

Index

Note: The bold numbers are the primary definitions/references.

A
absolute temperature **5**, 42
absolute zero **5**
adiabatic **46**, 47, 48, 127
   adiabatic, dry **49**, 186, 188
   adiabatic, moist **49**, 189
   adiabatic cooling **47**
   adiabatic warming **47**
adjustment 108, **113**, 263
advection **35**, 37, 85
   advection, cold **36**, 107
   advection, thermal **106**
   advection, warm **36**, 107, 197
ageostrophic wind **109**, 110–112, 155
albedo **251**
air mass **9–10**
amplitude **135**
anomaly **143**
anticyclonic **89**, 99, 107
atmospheric oscillations **203–205**
available potential energy **262**

B
balanced flow **95**, 95–103
baroclinic **14**, 247
   baroclinic instability **161–162**, 247–248
   baroclinicity **120**, 121, 127, 130
barotropic **14**, 120, 127, 140, 146
beta plane approximation **139**

bow echoes **197**
boundary layer **168**
Boussinesq approximation **104**, 149
Brunt-Väisälä frequency **49**
buoyancy force **48**, 74, 105

C
centrifugal force **67**, 68
centripetal force **67**, 69, 95
chinook 80, **227**
circulation **119**
   circulation, absolute **123**
   circulation, Bjerknes' theorem **120–121**
   circulation, indirect **255**
   circulation, Kelvin's theorem **119–120**
   circulation, relative **123**
   circulation, thermally direct **113**
closure **172**, 174
   closure assumptions **172**
cold core cyclone **16**
cold occlusion **161**
conditionally unstable **189**, 191, 205
confluence **112**, 155
continuity **51**
   continuity equation **52**, 104
continuum hypothesis **50**
convection **46**, 172, 189, 205, 248
convective available potential energy **190**
convective inhibition **190**

convergence **15**, 51, 53
covariance term **170**
Coriolis force **69–71**
Coriolis parameter **84**
critical wind speed **220**
cyclonic **89**, 99, 107
cyclogenesis **153**
cyclolysis **153**
cyclostrophic **97**

D
Dalton's Law **43**
dew point depression **11–12**
diffluence **112**, 155
dimensional homogeneity **79**
dispersion relation **136**
divergence **15**, 51, 52–53
doldrums **256**
downslope windstorm **226**
drag **167**, 175
dryline **196**
dry static energy **262**
dynamics **4**
dynamic similarity **83**
d'Alembert's paradox **167**

E
eddy momentum flux **261**
eddy viscosity coefficient **176**
effective gravity **68**
Ekman boundary layer equations **177**
Ekman spiral **178**, 179
Ekman transport **179**
enhanced greenhouse effect **253**
entrainment **174**, 191
evanescent wave **221**
exact differential **32**
extra-tropical cyclone **3**
eye **205**, 208

F
fall line **234**
First Law of Thermodynamics **46**
flux **52**
  flux divergence **171**
  flux, momentum **261**
  flux, vorticity **124** (see also *vortex strength*)

flux Richardson number **173**
foehn **83**, 227
force **57**
Fram **176**
free atmosphere **167**
frequency **135**
front **9**, 10–12, 149–152 (see also *Margules' Model*)
  front, cold **11**
  front, occluded **11**, 16
  front, polar **15**, 149
  front, stationary **11**, 151
  front, warm **11**
frontal cyclone **14**, 152
frontal system **3**

G
geopotential **59**
  geopotential height **59**
geostrophic **87**, 88, 112–113
  geostrophic approximation **87**, 98
  geostrophic wind **87**
glaciation **188**
gradient wind **99**, 100
gravity **58–59**
greenhouse effect **252–253**
Greenwich Mean Time (GMT) **9**
group velocity **136**
gust front **193**

H
hodograph **96**
horse latitudes **256**
hydraulic jump **230**
hydrostatic approximation **86**
hydrostatic balance **63**, 86
hydrostatic pressure **73**
hypsometric equation **64**

I
ideal gas **42**
Ideal Gas Law **42**
incompressible **72**
inertial frame of reference **58**
inertial oscillation **96**
inertial term **81**
instant occlusion **161**
Intertropical Convergence Zone **205**

intrinsic frequency 214
inversion 66
inviscid 120
isallobaric wind 111
isobaric 12
isotropy 169
isotherm 14
isothermal layer 66

J
jet **154**, 155
   jet, low level **195**
   jet, polar **154**
   jet entrance **154**
   jet exit **154**
   jet maximum **154**, 256

K
katabatic force 238
   katabatic wind 91, 233
Kelvin waves 202
kinematic viscosity coefficient 63

L
lapse rate **75**, 189
   lapse rate, adiabatic **189**
   lapse rate, moist adiabatic **189**
lee wave 225
lifting condensation level 187
long waves **13**, 17, 154 (see also *planetary waves* and *Rossby wave*)

M
Margules' formula 151
Margules' Model 149–151
mesocyclone 194
mesoscale 194
mesoscale convective complexes 197
mesoscale convective systems 196
mid-latitude westerlies **14**, 255
mixed layer theory 175
mixing length 180
mixing ratio 44
moist static energy 261

N
neutrally stable 49
no-slip boundary condition 168

non-dispersive 136
non-inertial frame of reference 28
non-linear differential equation 84
normal stress 41
norwesters 227

O
occlusion 17, 161
   occlusion, bent back 161
   occlusion, cold 161
   occlusion, instant 161
   occlusion, warm 161
open wave cyclone 15
orography 209

P
parcel 35
phase 135
phase speed 136
planetary waves **13**, 82, 140–143 (see also *Rossby wave* and *long waves*)
polar low 247
polar vortex 90
positive vorticity advection maximum 160
potential function 33
potential vorticity **126**, 128
potentiotropic 127 (see also *adiabatic*)
pressure 5
   pressure gradient force 59
   pressure, atmospheric 6
   pressure, hydrostatic 73
   pressure, mean sea level **6–7**, 8
   pressure, partial 43
   pressure, perturbation 73
propagate 17
propagating wave 135

Q
quasi-biennial oscillation 203
quasi-geostrophic flow **108**, 131

R
radiosonde 10
retrograde 142
retrogress 149
reverse shear 248
Reynolds decomposition 169
rheology 42

Richardson number **173**
ridge **13**
Rossby number **81**
Rossby's formula **140**
Rossby wave **141** (see also *planetary waves* and *long waves*)

S
saturation **184**
saturation vapor pressure **184**
sensitive volume **50**
scale height **64**
Scorer parameter **223**
shear stress **41**
short wave **13**, 15, 149, 154, 248
similarity theory **181**
specific humidity **44**
specific volume **46**
standing wave **135**
state **42**
station model **7–8**
stratification **214**
subcritical **229**
supercritical **229**
supersaturated **189**
surface weather map **11**
synoptic **82**, 86

T
Taylor-Proudman theorem **104**
temperature
　temperature, advection **37**
　temperature, atmospheric **4–5**
　temperature, dew point **7**
　temperature, potential **46**, 49, 127
　temperature, virtual **45**
　temperature, wet bulb **192**
Teten's formula **184**
thermal steering principle **159–160**
thermal wind relationship **106**, 113, 259
thickness **65**
trade wind **123**, 205, 253
tropopause **66**, 105, 145
troposphere **4**
trough **13**
turbulence **168**
turbulent eddy **168**
turbulent flux divergence **171**

U
unit vector **23**
Universal Time (UTC) **9**

V
vapor pressure **44**
variable, intensive **42**
variable, state **42**
viscous **41**
viscous force **61–63**
viscous sub-layer **168**
vortex strength **124** (see also *flux, vorticity*)
vorticity **124**
vorticity equation **131**

W
wall cloud **195**
warm sector **16**, 18, 161
wavenumber **136**, 141, 142
well-mixed layer **174**
wind shear **104**, 106, 113

WILEY COPYRIGHT INFORMATION AND TERMS OF USE

CD supplement to *Amanda H. Lynch & John J. Cassano, Applied Atmospheric Dynamics*.

Copyright © 2006 John Wiley & Sons, Ltd.

Published by John Wiley & Sons, Ltd., The Atrium, Southern Gate, Chichester, West Sussex, PO19 8SQ. All rights-reserved.

All material contained herein is protected by copyright, whether or not a copyright notice appears on the particular screen where the material is displayed. No part of the material may be reproduced or transmitted in any form or by any means, or stored in a computer for retrieval purposes or otherwise, without written permission from Wiley, unless this is expressly permitted in a copyright notice or usage statement accompanying the materials. Requests for permission to store or reproduce material for any purpose, or to distribute it on a network, should be addressed to the Permissions Department, John Wiley & Sons, Ltd., The Atrium, Southern Gate, Chichester, West Sussex, PO19 8SQ, UK; fax +44 (0) 1243 770571; Email permreq@wiley.co.uk.

Neither the author nor John Wiley & Sons, Ltd. Accept any responsibility or liability for loss or damage occasioned to any person or property through using materials, instructions, methods or ideas contained herein, or acting or refraining from acting as a result of such use. The author and Publisher expressly disclaim all implied warranties, including merchantability or fitness for any particular purpose. There will be no duty on the author or Publisher to correct any errors or defects in the software.